American University Studies

Series XIV
Education

Vol. 3

PETER LANG
New York · Berne · Frankfort on the Main · Nancy

Ralph E. Martin, Jr.

The Credibility Principle and Teacher Attitudes toward Science

PETER LANG
New York · Berne · Frankfort on the Main · Nancy

The Credibility Principle and Teacher Attitudes toward Science

Library of Congress Cataloging in Publication Data

Martin, Ralph E., 1951–
   The credibility principle and teacher attitudes toward science.
   (American University Studies. Series XIV, Education; vol. 3)
    Bibliography: p.
    1. Science – Study and teaching (Elementary) – Teacher training. 2. Teachers college students – Attitudes – Evaluation. 3. Attitude change – Evaluation. I. Title. II. Series.
LB1585.M29   1984   372.3'5044   83-49429
ISBN 0-8204-0101-3

CIP-Kurztitelaufnahme der Deutschen Bibliothek

Martin, Ralph E.:
The credibility principle and teacher attitudes toward science / Ralph E. Martin, Jr. – New York; Berne; Frankfort on the Main; Nancy: Lang, 1984.
   (American University Studies: Ser. 14, Education; Vol. 3)
   ISBN 0-8204-0101-3

NE: American University Studies / 14

© Peter Lang Publishing, Inc., New York 1984

All rights reserved.
Reprint or reproduction, even partially, in all forms such as microfilm, xerography, microfiche, microcard, offset prohibited.

Printed by Lang Druck, Inc., Liebefeld/Berne (Switzerland)

To Marilyn, Jennifer, and Jessica

TABLE OF CONTENTS

	Page
LIST OF TABLES	vii
LIST OF FIGURES	xiii

Chapter

I. THE PROBLEM 1

    Need for the Study 1

    The Problem 4

    Purpose of the Study 7

    Hypotheses 8

    Assumptions 9

    Limitations 10

    Definitions 10

    Overview 11

II. REVIEW OF RELATED LITERATURE 13

    Definition of an Attitude 13

    Importance of Attitudes to Teacher Education 15

    Learning Theory and Hovland's Principles of Attitude
        Change 17

    Effects of the Communicator on Attitude Change 22

    Nature of the Respondent and Attitude Change 32

    Summary 37

Chapter	Page
III. METHODOLOGY	41
Introduction	41
Population	42
Treatment	43
Instrumentation	47
Design and Data Collection	53
Hypotheses	55
Analyses	56
Summary	58
IV. ANALYSIS OF DATA	59
Introduction	59
Question 1	60
Question 2	67
Question 3	74
Question 4	77
Summary	93
V. CONCLUSIONS, IMPLICATIONS, AND RECOMMENDATIONS	101
Overview	101
Conclusions	101
Implications	110
Hovland's Credibility Principle	111
Science Education	112
Credibility Characteristics	112
Teacher Education	114
Recommendations	116
Concluding Statement	119

	Page
REFERENCES	121

APPENDIXES

A. The University of Toledo Competency - Based Teacher Education Program in Elementary Education 129

B. Elementary Teaching and Learning III, 312:328, Course Calendar, Winter Quarter, 1981 139

C. The University of Toledo Elementary Teaching and Learning III, 312:328, Module 02, Teaching Science In The Elementary School, Winter, 1981 Revision/Martin 143

D. Elementary Teaching and Learning III, 312:328, Course Evaluation Questionnaire 155

E. Science Teaching Attitude Scales (STAS) also known as What Is Your Attitude Toward Science and Science Teaching? by Richard Moore, 1973 169

F. Perceptions of Communicator Attitudes (PCA) 177

G. Perception of Communicator Credibility 183

LIST OF TABLES

Table		Page
1.	Science Attitude Statistics	61
2.	Science Teaching Attitude Statistics	63
3.	Descriptive Statistics of Posttest Perceptions of Communicator Credibility	69
4.	Posttest Population Perceptions of Communicator Credibility Rank	71
5.	Descriptive Statistics of Posttest Perceptions of Communicator Credibility	72
6.	Postposttest Population Perceptions of Communicator Credibility Rank	73
7.	Posttest-Postposttest Changes in Communicator Credibility	73
8.	Science Attitude Changes and Perceptions of Communicator Attitudes	76
9.	Science Teaching Attitude Changes and Perceptions of Communicator Attitudes	78
10.	Comparison of Specialization Group and PCA Science Attitudes	80
11.	Comparison of Specialization Group and PCA Science Teaching Attitudes	80
12.	Specialization Group Science Attitude Profiles	81
13.	Specialization Group Science Teaching Attitude Profiles	82
14.	Population Changes in Perception of Science Training From Pretest to Posttest to Postposttest	84
15.	Comparison of Science Training Perception Groups and Science Attitudes	86
16.	Comparison of Science Training Perception Groups and Science Teaching Attitudes	86

Table		Page
17.	Science Training Perception Group Science Attitude Profiles	87
18.	Science Training Perception Group Science Teaching Attitude Profiles	88
19.	Comparison of Achievement Group and PCA Science Attitudes	90
20.	Comparison of Achievement Group and PCA Science Teaching Attitudes	90
21.	Science Achievement Group Science Attitude Profiles	91
22.	Science Achievement Group Science Teaching Attitude Profiles	92
23.	Comparison of Completed Science Course Groups and PCA Science Attitudes	94
24.	Comparison of Completed Science Course Groups and PCA Science Teaching Attitudes	94
25.	Completed Science Course Group Science Attitude Profiles	95
26.	Completed Science Course Group Science Teaching Attitude Profiles	96

LIST OF FIGURES

Figures		Page
1	Research Design	54

Chapter I

THE PROBLEM

Need for the Study

Teacher attitude is one of the most important aspects of a school's effectiveness. Positive and negative teacher predispositions toward particular subjects contribute to the nature of a school's educational environment, determine how instructional resources are utilized in the classroom, and influence student attitudes and achievement. It is believed that the education of any one student for any one year is most dependent upon what that student's teacher believes, knows, and does -- and doesn't believe, doesn't know, and doesn't do. For essentially any subject learned in school, the teacher is the enabler, the inspiration and the constraint (Stake & Easly, 1978). Moreover, there is support for the position that a teacher's enthusiasm and relationships in the classroom are far more important than subject matter knowledge (Bybee, 1972; Shrigley, 1974). Also, teachers who possess positive attitudes toward a particular subject area and its teaching have been found to promote that subject in the classroom and cultivate similar positive attitudes among students (Washton, 1971).

Hence, two important factors emerge from the literature: (1) teacher attitudes do make a difference in the teaching-learning process (Good, Biddle & Brophy, 1975); and (2) teacher attitudes can be changed (Stern & Keislar, 1971). Both points spawned numerous education attitude studies during the past two decades. However, according to Shrigley (1976) many education attitude researchers repeated the mistake of the social psychologists in the 1930's when as Kiesler, Collins & Miller (1969, p. 8) reported, ". . . variables such as 'a year's course in economics' are of such gross and 'shotgun' nature that it is difficult to ascertain which aspect or facet of the complex manipulation is actually responsible for the results." Researchers who conducted studies of this design could not safely infer cause and effect unless extraneous variables were adequately controlled (Borg & Gall, 1979). Thus, social psychologists selected theories to provide a systematic framework for the identification of relevant and extraneous variables.

"A major role of theory," continued Kiesler (et al., 1969, p. 8) ". . . is to specify those aspects in a complex situation that contribute to attitude change." Once identified, the aspects of attitude change, as specified by the theory, can be tested for the purpose of declaring the theory reasonable or unreasonable in the special context of the attitude change environment. Such an approach is necessary because no single attitude change theory is universally accepted despite the complementary nature of many. Each theory has its limitations, different degrees of specificity, and areas of applicability (Oskamp, 1977).

In the special context of the teaching-learning process, a stimulus-response (S-R) learning theory may apply to attitude changes which occur in an education environment. Such a theory is one of several contemporary theories which supports the position of learned attitudes (Kiesler et al., 1969). Advocates of the S-R theory maintain that changes in attitudes are learned changes, that is, learned responses to stimuli, and are achieved under the same stimulus-response conditions as other learnings (Wagner & Sherwood, 1969).

Following the model of a stimulus-response learning theory, Carl Hovland, Irving Janis and Harold Kelley (1953) advanced attitude change principles of communication and persuasion. According to Hovland and his associates, in a stimulus-response environment for attitude change, the message communicator and the communication are considered the stimulus and the respondent's agreement with the message is the response (Shrigley, 1976). Among Hovland's principles, the message communicator was identified as the key agent of attitude change (Oskamp, 1977). The credibility of the communicator, as perceived by the respondents, was discovered to be a vital link between the attitude change message and the resultant degree of change in attitude (Kiesler et al., 1969). In such an environment, the communicator's primary method of communication is verbal persuasion (stimulus) to which the respondent's reward is varied forms of agreement (response) with the communication (Shrigley, 1976). Hovland and associates maintained that under these circumstances the perceived credibility of the communicator affected both the evaluation of the message received by the respondent and the amount of attitude change.

Hovland and his associates summarized the four basic components of the persuasive approach in a single statement: "Who says what to whom with what effect?" (Smith, Laswell & Casey, 1946). Hovland also suggested that communicators who are perceived as being highly credible and authoritative are more likely to produce greater attitude change, whereas communicators who are perceived to be less credible and authoritative are less likely to produce change. From these principles one may infer that the respondent's attitudes will move in a direction and approach the attitude level maintained by the more credible communicator.

The model suggested by Hovland's principles may be useful to attitude change studies in education. If one considers the classroom as the stimulus-response learning environment, the teacher as the communicator, and students as respondents to the teacher's message, then one may expect the more credible teacher (as perceived by the students) to effect a greater change in attitude than a teacher who is perceived as being less credible.

Hovland's attitude change principles of communication and persuasion have been proven useful in changing attitudes relevant to social issues (Hovland et al., 1953). The principles intuitively apply to education environments that desire to initiate changes in student attitudes. However, this researcher found no studies that tested these principles in a contemporary education context.

The Problem

Since teacher attitudes are an important factor in the teaching-learning process, teachers must have positive attitudes before entering the profession to ensure high levels of student attitudes and achievement. Therefore, attitude improvement is a worth-while goal of a preservice teacher's professional training (Morrisey, 1981).

One environment that might have impact on preservice teacher attitudes is the teacher education methods course where a credible communicator (instructor) can produce positive changes in attitudes. However, an important point must be noted about such a non-laboratory environment. Communicators do not work nor does communication take place in an antiseptic environment, wherein pure dosages in specific quantities can be administered to any respondent (Sherif & Sherif, 1969).

Numerous communicators (e.g., methods instructors, graduate assistants, university supervisors, field school cooperating teachers, and student peers) are commonly at work in campus-based teacher education methods classes and the accompanying practice teaching experiences in field schools. Each communicator may offer an attitude change message to the preservice teacher (respondent). Yet, each communicator cannot provide a communication free from the possible contamination of other communicators. The nature and quality of the message delivered by each communicator and the amount received by the respondent are not likely to be equal. Failure to account for differing perceptions of communicator credibility would limit the identification of attitude change elements in the methods course.

The nature of the respondents (preservice teachers) also must be considered to further the successful identification of attitude change elements. All preservice teachers enrolled in the methods course are not likely to compose a homogeneous group. Each preservice teacher may have encountered uniquely different experiences with the attitude object. Such experiences are of a direct personal nature and constitute the earliest and most fundamental way in which a respondent may form an attitude (Oskamp, 1977). Salient incidents, repeated exposures and stereotype images provide respondents with different perceptions of the usefulness of the attitude object and different perceptions of conditions under which the object was experienced (Oskamp, 1977; Loree, 1971).

However, many respondents may have similar perceptions of the attitude object due to the similarity of the conditions under which the object was experienced. For example, in a teacher-education program, students (respondents) may have experienced similar attitude object conditions because they chose to enroll in the same classes. Selection of classes is influenced by the students' chosen majors or areas of specialization. Thus, the self-selection process may cause students to become members of identifiable groups.

Groups to which respondents are exposed exert important influences on the attitudes possessed toward the person or object. School groups, followed by peer and reference groups, rank among the most important sources of pressure which influence attitude formation and change. Conformity pressures, intended or unintended, exerted by a group affect the perception of oneself relative to the advocated position of the group (Oskamp, 1977).

Sources of group influence in a teacher-education methods course may develop because of the respondent's specialization group (e.g., language arts, social studies, mathematics, science, early childhood, special education) and academic achievement. Both elements may influence the respondents' self-perceptions, professional preparation and teaching. Self-perceptions also influence the respondents' perceptions of the communicator's credibility, thus affecting the evaluation of the message and the degree of attitude change (Hovland et al., 1953).

One cannot attribute the total change in attitudes to the influence of a particular course and be satisfied that the course alone produced the change. To identify possible attitude change elements, research is needed which examines the credibility of each communicator and the nature of the respondents in relationship to the attitude measures. Hovland's simple statement: "Who says what to whom with what effect?" may help to guide an analysis of attitude change and to develop a more effective teacher-education methods course. Each of the four basic components must be considered in the analyses. In this study "who" was considered to be the communicators, "what" the attitude change communication, "whom" the preservice teacher, and "what effect" the change in attitudes.

Purpose of the Study

The influence of the communicator has been identified in the literature as the probable producer of most changes in attitudes. However, many different communicators exist in a teacher-education methods class with a related field experience component. Each communicator may influence, to some degree, the attitudes that preservice teachers have toward a given subject and its teaching. The purpose of this study was to test Hovland's principles of attitude change toward changes in science and science teaching attitudes in the context of an elementary preservice teacher science methods course. Therefore, the major question of concern to this study became one of:

> Does Hovland's principle of communicator credibility hold true in the context of a team taught preservice elementary teacher science methods course?

Given the context of this study, certain relationships must exist to confirm Hovland's principles. Therefore, to answer the major question, the following questions also had to first be answered:

1. Did preservice elementary teachers' attitudes toward science and science teaching change during and following the team taught, competency-based, field oriented science methods course?

2. Which communicator was perceived as being most credible during and following the methods course?

3. Did the preservice elementary teachers' attitudes change in a direction and approach the perceived attitude level of the most credible communicator?

4. Were the factors: specialization group, achievement in previous science courses, and self-perception of science training related to the preservice elementary teachers' attitudes toward science and science teaching?

Hypotheses

The following hypotheses were based upon the above questions:

1. Preservice elementary teachers who complete a science methods course with a field experience will demonstrate changes in attitudes during and following the methods course.

2. The perceived credibilities of the communicators are not equal.

3. Preservice elementary teacher attitudes change in the direction and approach the perceived attitude level of the most credible communicator.

4. A positive relationship exists between preservice elementary teacher attitudes and the perceived attitudes of the most credible communicator, within and conditioned by selected preservice elementary teacher factors.

The above hypotheses are restated in null form in Chapter III.

Assumptions

This study of the relationships among attitudes, communicator credibility, and selected respondent factors was based upon the following assumptions:

1. Attitudes are not innate, they are learned; therefore, positive attitudes can be taught (Shrigley, 1974).

2. Attitudes are influenced by a variety of variables which exist within the structure of a course (Haney, 1964).

3. The communication concerns science, various ways to teach it, its importance, and the role it has in the elementary curriculum.

4. Preservice elementary teachers are a valid population to determine the credibility of a communicator (Shrigley, 1976).

5. The preservice elementary teachers were honest and knowledgeable while responding to treatments and completing the instruments.

6. The instruments used to measure the dependent variables and methods of statistical analyses are valid.

7. An increase or decrease of at least three points is a meaningful difference in science and science teaching attitudes.

Limitations

The following items were limitations of the study:

1. The study was limited to an existing population of 25 preservice elementary teachers.

2. The study of attitudes required that qualitative terms be expressed in quantitative measures.

3. The course was team taught and included topics not directly associated with science methods, thus placing restrictions upon the design of the instructional methods module and the amount of instructional time provided the methods instructor.

4. The team teaching organization required a fixed calendar, thus limiting a concentration of science methods instruction.

Definitions

Attitude -- "A mental and neural state of readiness, organized through experience, exerting a directive or dynamic influence upon the individual's response to all objects and situations in which it is related" (Allport, 1935, p. 810).

Attitude Toward Science -- a quantified response to four positive and four negative Likert-type scales obtained for elementary teachers from Moore's (1973) "What is Your Attitude Toward Science and Science Teaching?" instrument.

Attitude Toward Science Teaching -- a quantified response to three positive and three negative Likert-type scales obtained for elementary teachers from Moore's (1973) "What is Your Attitude Toward Science and Science Teaching?" instrument.

Communicator -- usually an individual speaker who communicates directly to an audience and gives her own views on an issue (Hovland et al., 1953).

Credibility -- the expertness and trustworthiness of a communicator as perceived by the respondent (Hovland et al., 1953).

Respondent -- the recipient of the communicator's message who is required to express an attitude toward the communication (Hovland et al., 1953).

Overview

The remainder of the study is divided into four chapters. Chapter II contains a review of related literature dealing with selected principles of attitude formation and change, the effects of message communicators, and the various factors associated with the respondents. Chapter III contains a description of the population used, instruments employed, and the methodology of the study. Analyses of the data are provided in Chapter IV. Conclusions, implications, and recommendations for further research are contained in Chapter V.

Chapter II

REVIEW OF RELATED LITERATURE

The purpose of this chapter is to review the literature on attitude change that is related to the role of the communicator and nature of the respondent which may apply to a teacher-education environment. Knowledge of what an attitude is, its importance to teaching and learning, and the theory from which Hovland's principles of communicator persuasion and respondent attitude change emerged are requisite to this purpose. The following review of related literature consists of five major areas:

1. Definition of an Attitude
2. Importance of Attitudes to Teacher-Education
3. Learning Theory and Hovland's Principles of Attitude Change
4. Effects of the Communicator on Attitude Change
5. Nature of the Respondent and Attitude Change

Definition of an Attitude

Originally the term "attitude" referred to a person's body position or posture. In the social sciences attitude has come to mean a posture of the mind rather than posture of the body (Oskamp, 1977). Many definitions of attitude exist, each with varying emphases of the components believed to contribute to the hypothetical construct commonly called an attitude.

Gordon Allport's comprehensive definition of an attitude has been widely recognized by contemporary psychologists as perhaps the most complete definition (Oskamp, 1977). According to Allport (1935) an attitude is:

> . . . a mental and neural state of readiness, organized through experience, exerting a directive or dynamic influence upon the individual's response to all objects and situations in which it is related (p. 810).

J. D. Halloran (1970) identified three factors inherent to Allport's definition of an attitude. First, an attitude is believed to be a state of readiness which leads an individual to perceive objects and persons in certain ways. Persons are often ready to deal with the objects and people around them, as they are met, without having to stop and think about each encounter. Second, attitudes are not innate, they are learned. Attitudes develop and are organized through experiences. Moreover, the states of readiness are modifiable and subject to change through additional experiences. Third, attitudes are dynamic products of experience which act as directive factors when the individual enters into subsequent experiences. Hence, attitudes lead to evaluations which may be emotionally toned and cause one to order priorities between different responses and programs of actions. Finally, attitudes are not directly observable but must be inferred from other observable data.

If the three factors identified above apply throughout an individual's life to form what is commonly referred to as an attitude toward another person or object, it then follows that they also apply to an education context. Effects on perception and evaluation and the effects of experiences on the learning of attitudes affect teachers and students alike. The role of the teacher is particularly important because a teacher is an authority who has many opportunities to shape classroom curriculum and influence students.

Importance of Attitudes to Teacher Education

The 1960's and early 1970's witnessed the production of several elementary science programs. Each program recommended different styles of teaching and was developed from varying philosophies on the appropriate blend of content and process (Morrisey, 1981). Only a few programs survived and continue to be used in elementary schools. One reason frequently cited for this lack of survival is the teachers' attitudes toward science and elementary science teaching.

Program failures were criticized for their lack of attention to teacher attitudes (Morrisey, 1981). It is now evident that failure should not have occurred for that reason, because lack of teacher interest, that is, a lack of positive attitude, had been identified as a barrier to effective teaching during the early 1960's (Blackwood, 1964). Moreover, Ausubel (1964) promoted a widely recognized hypothesis about the importance of teachers being able to communicate a sense of excitement about the subjects they were teaching.

For a number of years following Ausubel's hypothesis, a teacher's attitude was believed to be related to pupil achievement. Evidence in the late 1960's confirmed that behaviors associated with teachers' attitudes toward the curriculum _were_ related to measures of pupil achievement (Rosenshine, 1970). Additional research during the 1970's continued to support that viewpoint (Rambally, 1977).

Teacher behavior was the link between teacher attitude and pupil achievement. Evidence provided by Hagerman (1974), Jingozian (1973), and Wish (1976) revealed that teacher attitudes were related to teaching behaviors. Those behaviors affected student attitudes toward the subject because teachers with positive attitudes promoted that subject more in the classroom. Promotion of the subject resulted in increased student attitudes. Increased student attitudes, in turn, resulted in higher levels of achievement.

Additionally, teacher attitudes are believed to underlie the personality characteristics of good and bad teachers. Support for the belief is offered by positive correlations found between good teachers, student achievement and student attitudes (Hamachek, 1969). Good teachers have also been found to teach in a way that reflected instructional flexibility, cognizance of student viewpoints and an appreciation of divergent student attitudes.

Teacher attitudes are an important factor in the teaching-learning process because they affect teacher behaviors, student attitudes and achievement (Good, Biddle, & Brophy, 1975). More attention should be paid to the attitudes of those training to become elementary teachers since elementary teachers must typically teach several different subject areas (Morrisey, 1981). Further, it is helpful to develop elementary methods courses that positively influence preservice teacher attitudes. It seems prudent to change attitudes before preservice teachers become inservice teachers.

It becomes important, then, to consider a method for changing existing attitudes for the betterment of the teaching-learning process.

A method for attitude change can be supplied by theory because a major role of a theory is to ". . . specify those aspects in a complex situation which contribute to attitude change" (Keisler, Collins & Miller, 1969, p. 8).

Learning Theory and Hovland's Principles of Attitude Change

A theory provides a systematic framework for the identification of variables. Once identified, aspects of attitude change, as specified by the theory, can be tested for the purpose of being declared reasonable or unreasonable in the special context of the attitude change environment.

Carl Hovland, Irving Janis and Harold Kelley (1953) studied several aspects of communication as a result of their Yale Communication Research Program. They described their program in terms of three principle characteristics: (1) theoretical issues and basic research; (2) theoretical developments from psychology and related fields; and (3) the testing of propositions by controlled experimentation in which specially constructed communications were presented under conditions where the effects of various factors influencing attitude change could be isolated.

Learning theory. Hovland and associates were explicit in stating that the purpose of their research was not one of presenting a systematic theory of persuasive communication. Instead, they chose to use characteristics of contemporary psychological learning theories to develop a framework of working assumptions about attitude change. Those assumptions were then used to test various aspects of persuasive communication.

Hovland et al. (1953, p. 3) based his assumptions of attitude change (later to be regarded as principles) upon a simplified stimulus-response ". . . learning theory developed by Hull and subsequently adapted to complex forms of social behavior by Millard and Dollard, Mowrer and others." Hovland et al. (1953, p. 3) considered Hull's learning theory useful because ". . . the implications of [Hull's] formulation for responses to symbols [were] particularly relevant to the study of communication."

The use of persuasive communication to cause changes in attitudes toward social issues, objects, and persons was of special interest to Hovland's research. Hovland and associates compared the process of attitude change to the learning of a habit or skill in a stimulus-response learning environment (Kiesler et al., 1969). In their persuasive communication research, the communicator and communication were considered aspects of the stimulus which were presented to the audience. The communicator's primary method of communication was verbal persuasion (the stimulus) to which the respondent's reward was varied forms of agreement (the response) with the position of the communicator's message (Shrigley, 1976).

The nature of persuasive communication. Hovland assumed that attitudes persisted until an individual encountered new learning experiences. He considered exposure to a persuasive communication, which induced an individual to form a new attitude, to constitute a learning experience.

> That is to say, when presented with a given question, the individual now thinks of and prefers the new answer suggested by the communication to the old [answer] held prior to exposure to the communication (Hovland et al., 1953, p. 10).

The "recommended position" presented in the communication was assumed to be "one key element." This element was considered a compound stimulus which raised the critical question and recommended a new answer.

When exposed to the recommended position, an individual was assumed to react with two distinct responses. First, an individual must think of her own position, then she must think of the position suggested by the communicator. Hence, according to Hovland et al., (1953, p. 11), ". . . a major effect of persuasive communication lies in stimulating an individual to think both of [her] initial [position] and of the new [position] recommended in the communication."

Merely thinking about the new position relevant to the old was not believed to lead to attitude change. Hovland assumed that acceptance was contingent upon incentives, and that attitude change depended upon a greater incentive for acceptance of the new position rather than maintenance of the old. A major basis for acceptance of a new attitude was assumed to be provided by arguments or reasons which, as perceived by individuals, constituted rational and logical support for the attitude change. In addition to supporting reasons, Hovland et al., (1953, p. 11) recognized the possible existence of ". . . special incentives involving anticipated rewards and punishments which motivate[d] the individual to accept or reject a given [attitude position]."

Hovland assumed that three main classes of stimuli were present in an attitude change environment and that they were capable of producing shifts in incentives. One class had to do with the observable characteristics of the communicator. A second class involved the environment in which the individual was exposed to the communication, including the reactions of other individuals who received the communication. A final class of stimuli included important contact elements such as arguments and appeals. Hovland believed the success of these incentives depended upon the predispositions of the individuals and that a successful communicator presented a communication in a way that was adapted to the verbal skill level of the individual and was capable of stimulating her motives to accept the recommended position.

Hovland also noted that several features existed which made the learning environment of mass communications different from academic teaching or from skill instruction. For example, when formal teaching normally occurred, the audience was ordinarily set to learn and recognized and accepted the status of students relative to the instructor.

Also, in many classrooms where verbal learning took place, a goal was to teach a large amount of information. To attain that type of objective, a great deal of practice was presumed necessary before the student could be expected to learn. In contrast, persuasive communication generally consisted of only a single statement, within the memory span of most individuals, which was repeated several times. The main difficulty for the communicator was to induce the audience to accept the communication message. When routine practice opportunities were limited, the communicator had to provide for transfer of training so the learner would apply the new attitude position to relevant situations.

In mass communications, the retention of newly acquired attitude positions sometimes suffered interference from subsequent practice of newer responses to the same stimuli. However, generally in formal instruction, as one acquired in schools, the individual was seldom exposed to competing instruction designed to break down the new attitude position which she had just acquired. In persuasive communications, "the audience [was] likely to be exposed to additional communications presenting completely different points of view and designed to create completely different [attitudes]" (Hovland et al., 1953, p. 17).

Hovland believed the special factors mentioned above applied to changing attitudes in mass communications, but they were not necessary for changing attitudes in formal education (e.g., classrooms). However, many changes in teaching techniques and teacher and student attitudes toward education have occurred since Hovland's research. These "special factors" now may be useful to attitude changes in education.

Hovland believed that the person or group perceived as originating the communication was an important factor which influenced the effectiveness of the communication. Perceptions were based upon cues provided as to the trustworthiness, intentions, and affiliations of the communicator. Hovland hypothesized that the perceived nature of the communicator would affect the way that the audience responded to auxiliary incentives (like appeals and arguments) present in the persuasive communications of daily life. Thus, Hovland and his associates investigated the effects of variations in the perceived trustworthiness and expertness of the communicator on the recipients' evaluation of the attitude change message and on their acceptance of the position advocated by the communication.

Effects of the Communicator on Attitude Change

> The effectiveness of a communication is commonly assumed to depend to a considerable extent upon who delivers it . . . . Approval of a statement by highly respected persons or organizations may have much the same positive effect as if they originate it (Hovland et al., 1953, p. 19).

Communication effectiveness. Differences in communication effectiveness may depend upon whether a person originates the message, endorses it, or the medium through which the message is transmitted. Despite these differences, Hovland believed the same basic factors and principles supported the operation of each of the many communication sources. Therefore, an analysis of the psychological processes causing the reaction to one kind of source were expected to be applicable to other sources.

A communicator was believed to affect attitude change in a variety of ways. For example, a charismatic, effective speaker might increase the likelihood of attention. A person of high status may increase the incentive value of the advocated attitude position. An expert, trustworthy individual may succeed through the use of arguments or appeals.

Hovland assumed that the various attitude changes attributed to a communicator resulted through audience perceptions (e.g., admiration, awe, fear, reward, punishment, trust, confidence, knowledge, intelligence, and sincerity). He also assumed that those perceptions and other attitudes which affected a communication's influence were learned by each individual under varied environmental conditions. Hovland et al., (1953, p. 20) believed that individuals acquired ". . . expectations about the validity of various sources of information and learn[ed] that following the suggestions of certain persons [was] highly reward[ed] whereas accepting what others recommended [was] less so." These products of learning constituted a complex set of attitudes, generalized to a wide variety and types of communication, and affected the individual's reactions to the communicator's messages.

Hovland conceded the variability of the conditions under which these attitudes toward communicators were learned; recognized certain kinds of attitudes as being important in all societies; and, hypothesized that general principles concerning antecedents and consequences of such attitudes might be expected to have a high degree of generality within our own culture. Therefore, Hovland and his associates chose to limit their research to the factors related to the credibility of the communication source. Their analysis of credibility focused on two problems. "How do differences in the credibility of the communicator affect (1) the way in which the content and presentation are perceived and evaluated? (2) the degree to which attitudes and beliefs are modified?" (Hovland et al., 1953, p. 21). Brief consideration of possible psychological processes which underlie communication effects and attitude changes were prerequisite to Hovland's research.

Psychological processes and communicator credibility. An individual's acceptance of a communicator's message partially depends upon how well informed and intelligent the communicator is judged to be. Despite the "expertness" of the communicator, an individual may still reject the message if the communicator is motivated to make

nonvalid assertions. Hovland et al. (1953) therefore believed it
necessary to

> . . . make a distinction between (1) the extent to which a
> communicator [was] perceived to be a source of valid assertions
> ('expertness') and (2) the degree of confidence in the communi-
> cator's intent to communicate the assertions he consider[ed]
> most valid ('trustworthiness') (p. 21).

Hovland referred to the resultant value (i.e., expertness and trustworthiness) as the "credibility" of the communicator.

Since little research existed investigating the effects of
expertness on attitude change, Hovland relied upon the suggestive
results of several studies. Bowden, Caldwell and West (Hovland et al.,
1953), using subjects from a broad age range,

> . . . determined the amount of agreement with economic position
> statements attributed to persons of different professions
> (e.g., lawyers, engineers, educators, etc.). The statements
> were approved most frequently when attributed to educators
> and businessmen, and least frequently when attributed to
> ministers (p. 22).

A study by Kulp provided evidence which revealed that, for graduate
students in education, the social and political opinions of professional
educators and social scientists were more influential than lay citizens
(Hovland et al., 1953, p. 22). Hovland conceded that additional factors
may have influenced the results of the studies, but also believed
it likely that the results were partly due to differences in the
perceived expertness of the communicator.

Numerous speculations existed about the effects of trustworthiness on attitude change. One general hypothesis was that when
a person was perceived as having definite intentions to persuade
others, it was likely that person had something to gain, and
was perceived as being less worthy of trust. Merton's analysis of

Kate Smith's war bond selling campaign, wherein she broadcast continually for eighteen hours, provided suggestive evidence on the importance of communicator sincerity (Hovland et al., 1953, p. 23). The presumed stress and strain of Ms. Smith's marathon broadcasts appeared to validate her sincerity and establish her trustworthiness.

Additional effects of audience perceptions of trust or distrust on attitude change were suggested by correlational data from Hovland, Lumsdaine, and Sheffield (Hovland et al., 1953, p. 24). The data involved audience reactions to open-ended questions for the purpose of showing the War Department's orientation films. Although the correlation may have indicated a general reaction to film content, Hovland recognized an alternative interpretation, that is, there existed a tendency to reject communications which were perceived as having manipulative intentions.

The research suggested that perceptions of communicator expertness and trustworthiness might affect an individual's attitude change. Hovland believed that evidence from his research would identify conditions under which the phenomenon occurred and provide insight as to the specific processes involved. Therefore, Hovland and his associates sought to answer two primary questions: How do variations in communicator credibility (expertness and trustworthiness) affect (1) the recipient's evaluation of the message, and (2) the amount of the recipient's attitude change?

How do variations in communicator credibility affect the recipient's evaluation of the message? Hovland reported the results of three separate studies concerning communicator credibility.

A study by Hovland and Weiss investigated the effects of credibility through the use of communicators differing in trustworthiness (Hovland et al., 1953, p. 27). Identical communications were presented to two groups of college students enrolled in an advanced undergraduate course. One group received a communication from a high credibility source, the other group from one of low credibility. Opinion questionnaires were given before, immediately following, and one month after the communication. Each subject responded to four contemporary topics based upon their reading of one article for each topic with the name of the source given at the end of the article. An analysis of credibility judgments obtained before the communications revealed that the sources used differed greatly in their credibility. Even though the communications were judged identical in content, ". . . the [articles] were considered to be 'less fair' and the conclusions 'less justified' when the source was of low rather than high credibility" (Hovland et al., 1953, p. 28). Hovland and Weiss concluded that judgments of content characteristics were significantly affected by variations in the credibility of the source.

Kelman and Hovland conducted another study which involved variations in source credibility (Hovland et al., 1953, p. 31). Two hundred and seventy-two seniors in summer school were divided into three groups and required to listen to and judge the educational value of a recorded radio program. A speaker was introduced during each program who gave a talk favoring extreme leniency in the treatment of juvenile delinquents. Three different introductory versions were used, one for each group. In the positive version, the speaker was introduced as a judge in a juvenile court. The speaker was presented as being highly trained, well-informed, experienced, and authority on criminology and delinquency. Also he was sincere, honest, and had the public's interests at heart. A neutral version presented the speaker as a randomly selected member of the studio audience. No additional information was provided. The speaker in the negative version was presented as being selected from the studio audience, but additional information emerged in the interview. The speaker had been delinquent as a youth and was involved in shady transactions as an adult. During the interview a low regard for the law, disrespect for parents, self-centeredness and a strong favor of leniency for delinquents was expressed.

Attitude data were obtained by an adapted scale on attitudes toward the treatment of criminals. The scale was administered after the radio program and again three weeks later. A special attitude measure established the equivalency of group attitudes before the experiment. Kelman and Hovland concluded (Hovland et al., 1953):

> With identical content, [student] judgments concerning the fairness of the presentation were much more favorable when it [program] was given by the positive communicator than by the negative one. The judgments for the neutral communicator were intermediate but more similar to those for the positive one (p. 32).

A study by Hovland and Mandrell concerned the variable of trustworthiness (Hovland et al., 1953, p. 33). Two hundred and thirty-five college students in introductory psychology classes were given a communication on the topic "Devaluation of Currency." The verbal communication contained an introduction which either (1) aroused suspicion of the communicator's motives, or (2) established belief in the communicator's impartiality. Subjects who received the suspicion arousing variation were informed that the speaker

was the head of a large importing firm who would profit from devaluation.
In the nonsuspicion variation, the speaker was introduced as an economist
from a reputable American university. The subjects were asked to
give their reactions to the program and speaker at a later time.

Data revealed that subjects perceived the "motivated" speaker
(suspicion arousing) as having done ". . . a poorer job and as having
been less 'fair and honest' in his presentation than the 'impartial'
communicator" (Hovland et al., 1953, p. 34). This occurred despite
the fact that the two speakers presented speeches that contained
identical content and conclusions. Hovland and Mandrell concluded
that cues as to the speaker's motives influenced the students' judgments
of the presentation and content.

How do variations in communicator credibility affect the
recipient's attitude change? In the study conducted by Hovland and
Weiss the subjects' attitudes changed in the direction advocated by
the communication more frequently when the message originated from
a high credibility source rather than a low credibility source
(Hovland et al., 1953, p. 29). However, when data were obtained
four weeks after the subjects read the articles, the effectiveness
of the communicators, with respect to high or low credibility, had
disappeared. Hovland and Weiss believed this finding occurred because
of a decrease in acceptance of the message advocated by high credibility sources. They also noted that the result could have been
attributed to forgetting of content, decreased awareness of credibility
due to the passage of time, or both.

An increase in attitude was demonstrated by the low credibility
group after four weeks. Hovland believed this occurrence suggested

that the negative effects of low credibility (i.e., untrustworthy) wore off and permitted the communication arguments to produce a delayed positive effect. According to Hovland's et al. (1953, p. 30) explanation, ". . . the effect of the source is maximal at the time of the communication but decreases with the passage of time more rapidly than the effects of the content." This explanation contributed to a general communciation principle posited by Hovland et al. (1953):

> . . . the stronger the perceptual response to the source during initial exposure to the communication, the more likely it is that the source will be [remembered] when, on subsequent occasions, any aspect of the communication is present. Strong responses to a communicator would presumably occur when the communication situation highlights [a communicator's] uniqueness as a source or when the situation forces the audience to consider [the source's] characteristics in evaluating the assertion (p. 31).

In the study conducted by Kelman and Hovland, the attitude changes closely paralleled the evaluations of the presentations (Hovland et al., 1953, p. 32). The group that heard the communication from the positive source favored more lenient treatment of delinquents than the group which heard it from the negative source. The attitude scores of the neutral source group were similar to those of the positive source. Kelman and Hovland interpreted those results to suggest that group perceptions of fairness and trustworthiness affected attitude change more than perceptions of expertness.

An alternative version of the attitude instrument was used three weeks after the communication. Kelman and Hovland reported no attitude differences among the groups and likened the results to those of the Hovland and Weiss study.

Results of the study conducted by Hovland and Mandell indicated no greater net gain in attitudes when the communication emanated from the suspicion arousing or non-suspicion communicator (Hovland et al.,

1953, p. 35). The differences in attitudes were not statistically significant (p = 0.23, one-tailed t-test), despite large differences in the subjects' evaluations of the two presentations.

Summary of the effects of communicator credibility. Hovland's research indicated that the recipients' reactions to a communicator were significantly affected by cues about the communicator's intentions, expertness, and trustworthiness. In all studies, the same communication was evaluated more favorably when presented by a communicator of high credibility. In two of the three studies, the most credible communicator evoked greater attitude change from the subjects. It was not possible for Hovland to "disentangle" the effects of trustworthiness and expertness -- the two main components of credibility. Both appeared to be important variables.

Discussion. Several interpretations may be offered to explain the fact that identical communications were evaluated differently by subjects exposed to sources of different credibility. One explanation was that people tended not to become exposed to communications from sources toward whom they had negative attitudes. However, all of Hovland's experiments involved captive audiences which could not easily resist exposure to the communications.

Hovland proposed two alternative explanations for the effectiveness of low credibility communicators:

1. Members of the audience did not pay close attention to the content nor did they attempt to comprehend its exact meaning because of their unfavorable attitudes. Instead, they spent

their time thinking about the communicator or "reading" meanings into the content message.

2. Members of the audience were not motivated to accept or believe the communicator because of their unfavorable attitudes.

In the studies reported by Hovland, little difference occurred between the amount of the content learned and the most and least credible communicator. Change of attitude did not only require learning, but also the motivation to accept the communicator's message. The critical aspect of attitude change was the degree to which the message recipients were motivated. Hovland et al. (1953, p. 38) reported that the evidence was quite clear, ". . . acceptance or rejection depend[ed] in part upon attitudinal reactions toward the source of the communication." The recipient was motivated to accept the attitude change messages which were anticipated to be reinforced by further experiences. Hovland reported that those anticipations were increased when a recommendation was presented by a more credible communicator and were decreased when cues of low credibility were present.

Hovland's research may be applicable to educational environments that desire to improve the teaching-learning process. Teacher attitudes influence teacher behaviors, and teacher behaviors affect student attitudes and achievement. Therefore, it is important to ensure that teachers complete their professional training with positive attitudes toward the subject areas they will be required to teach.

As suggested by Hovland's research, a credible university teacher (communicator), one who is perceived as being an expert, trustworthy authority figure, is more likely to present a message

which will persuade preservice teachers to change their attitudes
toward the advocated attitude position. Communicator credibility,
as perceived by the preservice teacher, will affect the evaluation
of the message. More credible communicators are likely to produce
more attitude change because of favorable evaluation and acceptance
of the message. Less credible communicators may undermine a preservice
teacher's attention to the content of the message, thus resulting in
a less favorable evaluation and less attitude change. However, in
a learning environment the effects of a communicator are determined
not only by the credibility of the teacher and the nature of the
message, but also by the motives and abilities of the students
(Hovland et al., 1953).

Nature of the Respondent and Attitude Change

The earliest and most fundamental way in which a student
learns and forms an attitude about an object or person is through
direct personal experience (Oskamp, 1977). Each student may have
encountered uniquely different experiences before entering and while
in the classroom; therefore, all students may not respond in the same
way to the same stimulus. It is necessary to give some consideration
to the differences the students may have toward the attitude object.

Experiences and predispositions. Salient incidents, particular
dramatic or traumatic events, experienced by children are believed
to influence attitudes, both immediate and those maintained later
in life (Fishbein & Ajzen, 1975). Such incidents can contribute to
predispositions which students may have toward the communication and
influence its effectiveness (Hovland et al., 1953). Stereotypes,
images or beliefs by a person about most members of a social group

or objects of a particular nature, are also known to influence attitudes (Oskamp, 1977). Stereotypes may be useful or they may result in false classifications due to incorrect images or beliefs held by a person. The effects of dramatic or traumatic events and stereotypes may, however, be counterbalanced by the cumulative effect of additional experiences (Oskamp, 1977). Repeated exposure to an object, person or idea over time is sufficient to enhance a person's attitude. Therefore, one may assume that adverse predispositions resulting from salient incidents or stereotypes can be overcome through credible communicators and positive, repeated experiences.

Achievement. Students can experience salient incidents in the classroom and form stereotype images as a result of the teaching-learning process. Such experiences and images may lead to preconceived notions about a subject area in which a student's attitude manifests itself in a particular level of achievement. Unfavorable perceptions of a teacher and the accompanying message are believed to be the result of unfavorable preconceptions (Hovland et al., 1953). Furthermore, perceptions are believed to obstruct the process of effective communication (Peterson & Carlson, 1974) and may result in reduced opportunities for achievement. A positive relationship usually exists between student attitudes and achievement in a particular subject, with students obtaining higher levels of achievement in subject areas where attitudes are high.

Hovland suggested that the reason for this relationship resulted because of differences in mental ability. Students with higher intellectual ability changed attitudes more rapidly because they recognized authoritative, cogent arguments and developed appropriate inferences based upon the logic of the communication. Hovland also noted that students of higher intellectual ability were likely to view arguments more critically, recognize a lack of expertness and trustworthiness, and change attitudes less when the communication was presented by less credible persons (Hovland et al., 1953). Therefore, a study of the relationship between the preservice teachers' attitudes toward a specific subject and perceptions of communicator credibility also should consider the possible effects of intelligence. An indicator of intelligence may be the preservice teachers' levels of achievement in the subject area of desired attitude change.

Self-perceptions. Preservice teachers' direct experiences gained through teacher education field experiences will most likely result in some self-perceptions of the quality of their professional preparation and their ability to teach a given subject. A resultant self-perception would be an ". . . attitude of the teacher toward herself" (Loree, 1971, p. 100).

An expectation of a teacher education program is to lead a teacher toward a ". . . better understanding of her own assets, beliefs, and values, and to help the teacher steadily improve her competencies" (Loree, 1971, p. 100). Persons with low self-perceptions are more likely to be influenced by persuasive communications and credible communicators (Hovland et al., 1953). Therefore, the perceptions of an individual's professional preparation may assist or obstruct the process of effective communication (Peterson et al., 1974) during the methods instruction of a teacher's training. Self-perceptions of training may have some bearing on the amount of the attitude change message accepted and the degree of success experienced by a teacher during the field experience portion of her training. Self-perceptions also should be considered when examining the relationship between the preservice teachers' attitudes and perceptions of communicator credibility.

Group membership. In addition to achievement and self-perception, the various groups to which preservice teachers belong are believed to exert important influences on their attitudes. Schools, second only to parents, exert important influences on student attitudes and become an even more important source of influence as the student matures (Oskamp, 1977). Highly favorable attitudes toward school, academic subjects, and professionals largely are due to the influence of classroom teaching (Oskamp, 1977). Also, friends' attitudes and intentions constitute conformity pressures that have been found to influence an individual's behavior and attitudes (Coleman, 1961).

A variety of other conformity pressures also can lead to attitude formation and change. Cooperative efforts toward common goals and intergroup contact, under proper conditions, have diminished prejudice and exerted powerful influences on attitudes (Crooks, 1970). Reference groups appear to provide a more mild influence on attitudes, often unintentional. These groups provide standards against which an individual may be measured. The central point is that reference groups may influence an individual's attitudes without attempting to do so and without an individual's knowledge (Oskamp, 1977). Influence is largely the result of predispositions upon which the group standards are based (Hovland et al., 1953).

Some of the most important of these predispositions have to do with the conformity to group attitudes resulting from membership in the group. The tendency of an individual to conform to group norms is based upon ". . . knowledge of what behavior is expected by the other members and on motives to live up to the [group's] expectations" (Hovland et al., 1953, p. 134).

An analysis of the influence of groups upon the attitudes of their members was of importance to Hovland's general problem of changing attitudes through communications. It is likely that group membership could also exert influences on attitude changes among preservice teachers due to resistance fostered by predispositions that the groups, to which the teachers belong, have toward certain communicators and the messages presented. A source of possible predispositions may result through preservice teacher memberships in groups that are a function of their academic programs.

It has been established that ". . . those who frequently interact with one another [and] take part in common activities . . . are generally considered as 'belonging together'" (Hovland et al., 1953, p. 135), that is, they form a group. Among preservice elementary teachers, groups form due to their chosen areas of specialization (e.g., language arts, mathematics, science, social studies, early childhood and special education). Within these groups, the preservice teachers are exposed to communicators who deliver messages which are quite likely to be of high interest by virtue of self-selection. Under these circumstances, resistance to change, as suggested by "foreign" communicators, is likely to be high (Hovland et al., 1953). When the preservice teachers are subjected to communication outside their areas of specialization, interests in the message may be quite low, their motives to accept the message low, preconceptions may rule, and resistance to change may be increased (Hovland et al., 1953). Therefore, preservice teachers' specialization groups also may have some influence on attitude change and perceptions of communicator credibility. These groups should be considered when examining the relationship between attitudes and communicator credibility.

Summary

Attitudes maintained by university teachers and teacher-education students are important to the successful preparation of future teachers. Teacher attitudes have been found to affect teacher classroom behaviors. These teaching behaviors also have been reported to influence student attitudes toward and achievement in specific subject areas.

To ensure positive student attitudes and high levels of achievement, it is prudent for teacher educators to promote the development of positive attitudes among preservice teachers. Positive attitudes are essential toward the subject areas they will be required to teach if preservice teachers are to become successful inservice teachers. Hovland, Janis and Kelley's (1953) principles of communication and persuasion may provide useful working principles for attitude change within the dynamics of a teacher-education program. Their principles, as they may relate to teacher education, include:

1. Trustworthiness and expertness are the main components of a communicator's credibility.

2. High credibility communicators have a greater affect on student attitude change than low credibility sources.

3. Communications attributed to low credibility sources are considered more biased and unfair than identical ones attributed to high credibility sources.

4. Effects on attitudes are not the result of differences in the amount of student attention or comprehension. Effects are related to the student's motivation to accept the attitude change message offered by the high credibility communicator.

5. The positive effect of high credibility and negative effect of low credibility communicators tend to disappear after a period of several weeks.

6. Some of the strongest predispositions are those related to the groups in which the student holds membership.

7. Differences in mental ability may affect the extent to which a student is susceptible to communication. Students with higher intellectual ability change more rapidly because they recognize cogent arguments and develop appropriate inferences, but they also view arguments more critically when presented by less credible persons.

8. Persons with low self-perceptions are more likely to be influenced by persuasive communications and credible communicators.

9. Greater attitude change occurs through active, rather than passive, student participation.

Hovland, Janis, and Kelley's principles worked well when the number of communicators was limited to three for short, intense periods of communication exposure. It has only been assumed that the principles apply to teacher education. It was not known if these principles applied to team taught courses where many communicators participated in the instruction. Therefore, the major question of concern to this study became: Does Hovland's communicator credibility principle of attitude change hold true in the context of a team taught preservice science methods course?

Chapter III

METHODOLOGY

Introduction

According to Hovland, Janis and Kelley (1953), the message communicator is the vital link between the communication and attitude change. Respondent perceptions of communicator credibility (i.e., expertness and trustworthiness) influence the evaluation of the communication and affect the direction and degree of attitude change. High credibility communicators have been found to have a greater affect than low credibility sources.

However, many different communicators exist in a teacher education methods course. College instructors and graduate assistants are primary sources of campus-based communication, while university supervisors and classroom cooperating teachers offer messages to preservice teachers during field-based experiences. Preservice teacher peers, particularly team members, are sources of communication in both environments. Each communicator may influence, to some degree, the attitudes that preservice teachers have toward a given subject and its teaching. Thus, the purpose of this study was to test Hovland's principle of communicator credibility on attitude change in the context of an elementary preservice teacher science methods course. The major question of concern to this study became:

Does Hovland's attitude change principle of communicator credibility hold true in the context of a team taught preservice elementary teacher science methods course?

Given the context of this study, certain relationships had to exist to confirm Hovland's principles. Relationships were determined by answering the following questions:

1. Did preservice elementary teachers' attitudes toward science and science teaching change during and following the team taught, field-oriented science methods course?

2. Which communicator was perceived as being most credible during and following the methods course?

3. Did the preservice elementary teachers' attitudes change in a direction and approach the perceived attitude level of the most credible communicator?

4. Were the factors of specialization group, achievement in previous science courses, and self perceptions of science training related to the preservice elementary teachers' attitudes toward science and science teaching?

Population

A population of preservice elementary teachers of The University of Toledo's College of Education was available for the study. The population consisted of 25 junior and senior elementary education majors enrolled during the Winter Quarter, 1981 in the course "Elementary Teaching and Learning III," a science methods course.

The population was composed of 22 females and 3 males ranging in age from 19 to 40 with a mean age of 21.7 years. The number of preservice teachers representing six areas of program specialization were: Early Childhood - 7; Language Arts - 5; Special Education - 5; Mathematics - 3; Social Studies - 3; and, Science - 2.

All elementary preservice teachers must complete four science courses before graduation, one each in biology and geology and two in natural science. Seven individuals had completed or exceeded the requirement before enrolling in the science methods course. Ten preservice teachers were completing a required science course concurrently.

The preservice elementary teachers had completed two of four required block courses in the professional education sequence. The first two blocks contained mathematics and language arts methods instruction. "Elementary Teaching and Learning III," science methods, was the third course in the sequence. It is followed by instruction in social studies methods and a quarter of elementary student teaching. All individuals had previous preservice teaching experience in local elementary schools. The science methods course was the first block that required team teaching with peers. See Appendix A for a more detailed description of the elementary education professional blocks and related field experiences.

Treatment

The treatment of this study consisted of three parts: (1) on-campus methods instruction and science unit planning during weeks 1-7 of the quarter; (2) in-field preservice elementary teacher science instruction of students in local elementary schools during weeks 8-10; and (3) an on-campus science fair and evaluative feedback sessions during the final week of the quarter (see course calendar, Appendix B).

The independent variables included six communicators (science instructor, graduate assistant, unit evaluator, university supervisor, cooperating teacher and peer team member), communicator communications and selected preservice teacher factors (area of specialization, perception of science training, and science achievement). Dependent variables included the preservice teachers' science and science teaching attitudes, perceptions of communicator credibility, and perceptions of communicator attitudes. Interaction among the variables may be examined as suggested by Hovland et al. (1953), "Who says what to whom with what effect?"

<u>Methods instruction and unit planning</u>. During methods instruction and unit planning, the primary communicators ("who") consisted of science instructor, graduate assistant and unit evaluators. Communications ("what") were related to elementary science teaching, the inclusion of generic instructional methods (see Appendix A, e.g., questioning, inquiry, problem-solving, concept attainment, etc.), and the development of unit plans for teaching a topic in science to elementary students. Preservice elementary teacher ("to whom") perceptions of communicator credibility and attitudes ("what effect") were based upon the communicator's expertness and trustworthiness in science and science teaching.

Science methods instruction consisted of $22\frac{1}{2}$ hours of experience with "hands-on" elementary science activities and discussions regarding science instruction, classroom management, and assigned readings (see Appendix C). The activities were selected from contemporary elementary science curricula, stressed both content and process, applied generic methods to science teaching, and were applicable to the preservice teachers' unit topics. In addition to science laboratory activities, preservice teachers were required to apply both science and general methods by team teaching a lesson to peers in a micro-teaching laboratory before implementing the lesson in the field. Objective feedback was offered by the science instructor, graduate assistant, and peers. The construction of learning centers, bulletin boards, and the presentation of projects in a science fair were also required.

The following is an account of the modular instruction and hours spent teaching by the team.

	Instructional Modules	Hours
*1.	Unit Planning and Implementation	13½
*2.	Teaching Science in the Elementary School	22½
*3.	Selection and Use of Instructional Media	3
*4.	Basics of Preparing Visual Instructional Materials	6
5.	Problem Solving	3
*6.	Classroom Management Techniques	1½
*7.	Concept Lessons	3
*8.	Inquiry Teaching	3
*9.	Questioning Lessons	1½
10.	Special Education	6
	Total	63

*These eight modules included a field experience which involved the use of checklists to assess the students' performances.

Field experience. University supervisors, classroom cooperating teachers, and peer team members were the principal communicators ("who") as the preservice teachers ("whom") spent 50 hours teaching science to students in local elementary schools. Thirteen preservice teachers taught in urban schools and 12 taught in a suburban school.

University supervisors and cooperating teachers supervised the teaching of science daily. All preservice teachers were required to evaluate their teaching, revise instruction, and meet with supervisors for daily seminars. Communication ("what") consisted of constructive criticism, suggestions for instructional improvement, and feedback from elementary students. Preservice teacher competencies in planning, methodology, and personal and professional fitness were evaluated by university supervisors using objective checklists (see Appendices A, C, and D). University supervisors shared the data with the preservice teachers during daily seminars and personal conferences. Perceptions of communicator credibility and attitudes ("what effect") resulted through daily interactions.

Science fair and feedback. All preservice teachers ("whom") assembled on campus for the science fair and evaluative feedback sessions ("what") during the last week of the quarter. Each preservice teacher displayed two projects from her teaching, received feedback from peers, science instructor, graduate assistant, unit evaluator and supervisor ("who"), and visited peer displays. Following the science fair, a slide/tape presentation depicted scenes of the preservice teachers' science laboratory and field experiences. The presentation was prepared by the science instructor and graduate assistant.

All preservice teachers completed a comprehensive evaluation of the course (see Appendix D). Results of the evaluation were shared with the preservice teachers during the next class day. Additional suggestions for course improvement were sought by the instructional team.

Instrumentation

The following are descriptions of instruments used to measure the dependent variables of the preservice elementary teachers': (1) science and science teaching attitudes; (2) perceptions of communicator attitudes toward science and science teaching; and (3) perceptions of communicator credibility.

<u>Science and science teaching attitudes</u>. Moore's (1973) <u>Science Teaching Attitude Scales</u> (STAS) instrument was used to determine preservice elementary teacher attitudes (Appendix E). Also known as "What is Your Attitude Toward Science and Science Teaching?" the instrument contains 70 items which assess attitudes in two areas: (1) attitudes toward science; and (2) attitudes toward elementary science teaching.

The Likert-type, forced choice instrument provided scores on 14 individual scales. Four scales assessed the following positive attitudes toward science:

1-P The laws and/or theories of science are approximations of truth and are subject to change.

2-P Observation of natural phenomena is the basis of scientific explanation. Science is limited in that it can only answer questions about natural phenomena and sometimes it is not able to do that.

3-P Science is an idea-generating activity. It is devoted to providing explanations of natural phenomena. Its value lies in its theoretical aspects.

4-P Progress in science requires public support in this age of science; therefore, the public can understand science and it ultimately benefits from scientific work.

Four scales assessed the following negative attitudes toward science:

1-N The laws and/or theories of science represent unchangeable truths discovered through science.

2-N The basis of scientific explanation is in authority. Science deals with all problems and it can provide correct answers to all questions.

3-N Science is a technology-developing activity. It is devoted to serving mankind. Its value lies in its practical use.

4-N Public understanding of science would contribute nothing to the advancement of science or to human welfare; therefore, the public has no need to understand the nature of science. They cannot understand it, and it does not affect them.

Three scales assessed the following positive attitudes toward science teaching:

5-P The idea of teaching science is attractive to me; I understand science and I can teach it.

6-P There are certain processes in science which children should know, i.e., children should know how to do certain things.

7-P Science teaching should be guiding or facilitating of learning. The teacher becomes a resource person.

Three scales assessed the following negative attitudes toward science teaching:

5-N I do not like the thought of teaching science.

6-N There are certain facts about science that children should know.

7-N Science teaching should be a matter of telling children what they are to learn.

Scores on each scale may be combined to provide positive and negative attitude scores for science and science teaching. Total attitude scores toward science and science teaching may also be derived from the instrument. Moreover, a universe total score may be computed by combining all scores on all scales. The universe total score may be used to describe combined attitudes toward science and science teaching. Total attitude scores toward science and total scores toward science teaching were used in this study.

The range of each scale is 0-15. The eight positive and negative attitudes toward science scales have a range of 0-120, whereas the six positive and negative attitudes toward science teaching scales have a range of 0-90. Universe total scores have a range of 0-210.

Moore (1973) established the reliability of the <u>Science Teaching Attitude Scales</u> by using the test-retest method. A reliability coefficient of 0.816 was obtained on the attitudes toward science scales and 0.934 on the attitudes toward science teaching scales. In a separate study, Riley (1979) reported Hoyt Reliability coefficients of 0.71 and 0.84 respectively on Moore's instrument.

The STAS instrument was field tested to demonstrate its construct validity. Moore (1972) designed the field test according to Kerlinger. Kerlinger (1964) proposed that:

> One can manipulate communications, for example, in order to change attitudes. If attitude scores change according to theoretical prediction, this would be evidence of the construct validity of the attitude measure, since the scores would probably not change according to the prediction if the measure were not measuring the construct (p. 451).

Thirty-one elementary school teachers participated in the field test. The STAS instrument was administered to the participants as a pre-pretest, pretest and posttest time series design. The hypotheses were:

1. There will be no significant difference between the pre-pretest and the pretest total scores on the STAS, and

2. There will be a significant increase from pretest to posttest on the total scores on the STAS.

Moore reported that acceptance of both of the hypotheses could be considered as evidence of construct validity.

The treatment given the participants was not designed to develop the attitudes assessed by the instrument. That is, the participants were not "propagandized" in order to produce changes on the STAS. However, positive changes from pretest to posttest were expected because the participants did receive intensive preparation for teaching Science Curriculum Improvement Study (SCIS) materials. Moore reported that the preparation was consistent with the attitudes assessed by the STAS.

Analysis of variance for repeated measures was used to test the significance between pre-pretest, pretest, and posttest means. The F ratio was significant beyond the 0.01 level. Analysis of variance was repeated for each scale. The Duncan multiple range test of significance for differences between the means was performed for the pre-pretest -- pretest and pretest -- posttest pairs of means according to a method described by Winer (1962).

Moore's data failed to reject the hypotheses. Therefore, Moore (1972, p. 15) concluded that "Acceptance of the hypothes[es] . . . [was] evidence of the construct validity of the [STAS]" since attitude scores changed according to the theoretical prediction.

Perceptions of communicator attitudes. Moore's (1973) STAS instrument was also used to obtain the preservice elementary teachers' perceptions of the communicators' attitudes (PCA) toward science and science teaching. The preservice teachers were directed to respond to each item on the STAS as they believed each communicator would respond if he or she were asked to complete the same instrument. Communicators included science instructor, graduate assistant, university supervisor, cooperating teacher and peer team member. The unit evaluator was excluded for economy of time. Because of the dual roles of some communicators (e.g., the graduate assistant and one university supervisor were also evaluators), the data were obtained from other communicator measurements. The preservice teachers recorded their perceptions of attitudes on answer sheets designed specifically for this purpose (see Appendix F).

Perceptions of communicator credibility. Preservice elementary teacher perceptions of communicator credibility (PCC) were measured by the Semantic Differential (SD) technique (Appendix G). Osgood's (1971) semantic differential has been used in a number of attitude studies (Butzow & Davis, 1975; Butzow & Williams, 1973; Gallagher & Korth, 1969, Klopfer, 1966). Since expressed perceptions are considered to be opinions and opinions are verbalizations of attitudes (Hovland et al., 1953), an attitude instrument such as the SD was considered a reliable instrument for measuring the concept of credibility.

The Semantic Differential has become a useful and powerful tool in quantifying the measurement of meaning. Osgood's SD is actually a scale in itself. The scale is reported to be of such a general sort that it can be applied to any concept at all (Oskamp, 1977) and can easily be modified to measure specific concepts (Triandis, 1971).

Unlike other types of instruments, the SD does not require a pool of statements. A listing of bipolar adjectives (e.g., good-bad, important-trivial, etc.), called scales, are used for each of many concepts to which individuals respond at one of several points on a continuum between each set. Based upon extensive research, Osgood (1971) reported the optimum number of points to be seven.

Osgood and his associates identified three factors which appeared among the scales. The factors, since labeled as dimensions, were evaluation, potency, and activity. The three dimensions accounted for nearly 50% of the total variance as measured by the SD technique. The technique requires the selection of a few scales from a factor study of each of the three dimensions to be used in the instrument. Three to four scales per dimension are sufficient to measure attitudes (Osgood, 1971).

Osgood used the evaluative dimension to measure attitudes because of its indication of affect. The potency and activity dimensions provided indications of the subjects' cognition. Osgood reported that in predicting Thurstone or Likert scales it was desirable to consider the information obtained from all three dimensional scales. Osgood also reported that multiple correlations predicting Thurstone scaled attitude scores from all three semantic differential dimensions were higher than the correlation between Thurstone scaled scores and the evaluative scores.

Jenkins, Wallace and Suci (1958) compiled an atlas of semantic profiles for 360 words. These words were rated on twenty scales by eighteen groups of thirty students. The reliability of the scale values was determined to be 0.97 for sophomore college students.

Mean scale values were found to correlate 0.97 with median score values. Osgood's study (1952) reported an overall reliability coefficient on the Semantic Differential of 0.85 for college students.

From the factor loadings reported in the atlas, the researcher selected thirteen pairs of bipolar adjectives believed to relate to the concepts to be measured. A panel of nine former elementary science methods students volunteered to participate in a field test of the instrument during early January 1981. The participants were asked to complete the instrument and during an interview respond to the clarity of the printed instructions and relationships between the bipolar adjective pairs and the concepts. As a result, the instructions were modified to the form indicated in Appendix G and four adjective pairs were eliminated: hard-soft, heavy-light, fast-slow, and excitable-calm.

Shaw and Wright (1967) reported that the attributes of the Semantic Differential appeared acceptable relative to other attitude scales. The test-retest method yielded reliability coefficients from 0.83 to 0.91. Evidence of validity has been estimated by correlations with Thurstone scales, 0.74 to 0.82. Guttman scale scores correlated 0.79 with scores obtained from a three-item Semantic Differential. Based upon extensive study, Osgood (1971) reported no reasons to question the validity of the SD on the basis of its correspondence with the results to be expected from common sense.

Design and Data Collection

The study was based upon a one-group pretest-posttest-postposttest research design (see Figure 1).

All preservice elementary teachers enrolled in the science methods course, "Elementary Teaching and Learning III," during Winter

Time line	Fall Quarter, 1980 Week 11	Winter Quarter, 1981 Week 1-7	Week 7	Week 8-10	Week 11
	Pretest	Treatment	Posttest	Treatment	Postposttest
	$STAS_1$	SMM	$STAS_2$	FE	SF,E $STAS_3$
			PCA_1		PCA_2
			PCC_1		PCC_2 SPST

$STAS_{1,2,3}$ - preservice teachers' attitudes toward science and science teaching.

SMM - Science Methods Module, generic methods and unit planning.

FE - Field Experience, subjects taught self-constructed science unit.

PCA_1 - perceptions of science instructor and graduate assistant attitudes toward science and science teaching.

PCA_2 - perceptions of university supervisor, cooperating teacher, and peer team member attitudes toward science and science teaching.

$PCC_{1,2}$ - perceptions of all communicators' credibilities.

SF,E - Science Fair and course Evaluation.

SPST - self-perceptions of science training.

Figure 1. Research Design.

Quarter, 1981, were pretested on their attitudes toward science and science teaching at the end of the Fall Quarter, 1980. Attitudes were measured by Moore's (1973) "What is Your Attitude Toward Science and Science Teaching?" instrument (Science Teaching Attitudes Scales, STAS).

After applying the treatment during the quarter's first seven weeks, all preservice teachers were given the STAS posttest. At that time, all preservice teachers also provided perceptions of all communicators' credibilities by completing the Perception of Communicator Credibility (PCC) instrument. Perceptions of science methods instructor and graduate assistant attitudes were measured by the Perceptions of Communicator Attitudes (PCA) instrument.

Postposttest attitudes were measured after the preservice teachers taught their self-constructed science units to students in local elementary schools. Again, the preservice teachers provided perceptions of all communicators' credibilities by completing the PCC. Perceptions of university supervisor, cooperating teacher, and peer team member attitudes were measured by using the PCA.

Self-perceptions of science training were classified as high, above average, average, and low as measured by the 5-P and 5-N scales of the STAS postposttest.

Achievement in previous science courses, obtained from science grades available in student records, were classified as high, above average, average and low achievement. Preservice teacher areas of specialization were also obtained by an inspection of student records.

Hypotheses

The following hypotheses resulted from the questions presented in Chapter I:

Null Hypothesis 1. As measured by the Science Teaching Attitude Scales (STAS), no difference exists between preservice teacher attitudes toward science and science teaching during and following the team taught, field oriented elementary science methods course.

Null Hypothesis 2. As measured by the PCC, no difference exists in the preservice elementary teachers' perceptions of the communicators' credibilities.

Null Hypothesis 3. A comparison of measures from the STAS and PCA will reveal no relationship between the preservice elementary teachers' attitudes and the perceived attitudes of the most credible communicator.

Null Hypothesis 4. No relationship exists between preservice elementary teacher attitudes and the perceived attitudes of the most credible communicator respective to preservice teacher science achievement, perceptions of science training, and specialization area.

Analyses

Inferential statistics were judged inappropriate because an entire population was available for the study. Descriptive statistics were selected to describe the changes in preservice teacher attitudes and perceptions of communicator credibilities and attitudes.

Null Hypothesis 1. Data analysis for Null Hypothesis 1 consisted of comparing pre, post, and postposttest science and science teaching attitude scores as measured by Moore's (1973) Science Teaching Attitude Scales (STAS). A SPSS computer program analyzed preservice teacher attitudes and reported data for each individual on each of the fourteen scales. Positive, negative, and total scale scores were provided for science and science teaching attitudes for each individual. Population pre, post, and postposttest attitude means were computed. Difference scores were also computed and reported in tabular form for each individual. Differences between the population means and the direction and amount of attitude change from pre, post, and postposttest for each individual were compared for the purpose of testing the null hypothesis.

Null Hypothesis 2. The perceived credibility of each communicator was measured by the PCC instrument at posttest and postposttest times. The concepts of expertness and trustworthiness were measured separately and combined into a single credibility score. Communicators were ranked from high to low in terms of credibility on the basis of the population means. Since some different communicators received equal credibility ratings from individuals, the most credible communicators were identified for specific preservice teachers from a frequency distribution. Changes in credibility perceptions, direction, and the degree of change were compared to test the null hypothesis.

Null Hypothesis 3. Preservice teacher pre, post, and postposttest STAS scores were compared to the perceived attitude scores (PCA) of the communicator(s) judged most credible. Individual preservice teacher profiles of attitude changes were constructed to test the null hypothesis.

Null Hypothesis 4. Individuals were divided into groups based upon specialization areas, perceptions of science training, and achievement in science. To test Null Hypothesis 4, attitude change profiles were constructed for the individuals within the group and relationships were examined according to the procedures listed for Null Hypothesis 3.

Summary

The message communicator has been identified as the vital link between the communication and attitude change (Hovland et al., 1953). However, many different communicators exist in a team taught teacher-education methods course. Therefore, the major purpose of this study was to test Hovland's principle of communicator credibility in the context of a team taught preservice elementary teacher science methods course.

To confirm Hovland's principles within this context, certain relationships had to exist including: (1) attitude changes during the course and field experience; (2) identification of the most credible communicator; (3) individual attitude changes in a direction and approaching the attitude level of the most credible communicator; and, (4) determining whether or not selected individual factors were related to attitude changes. Analyses of the data are presented in Chapter IV.

Chapter IV

ANALYSIS OF DATA

Introduction

The purpose of this study was to test Hovland's principle of communicator credibility on preservice elementary teacher attitude changes. The major question to be answered was:

Does Hovland's principle of communicator credibility hold true in the context of a team taught preservice elementary teacher science methods course?

Given the context of this study, certain relationships had to exist to confirm Hovland's principles. These relationships were examined using the following questions:

1. Did preservice elementary teachers' attitudes toward science and science teaching change during and following the team taught, field oriented science methods course?

2. Which communicator was perceived as being most credible during and following the methods course?

3. Did the preservice elementary teachers' attitudes change in a direction and approach the perceived attitude level of the most credible communicator?

4. Were the factors of specialization group, achievement in previous science courses, and self-perceptions of science training related to the preservice elementary teachers' attitudes toward science and science teaching?

The following analyses are related to each of the previous questions. Answers to each question are provided with supportive data.

Question 1

Did preservice elementary teachers' attitudes toward science and science teaching change during and following the team taught, field oriented science methods course?

Attitudes toward science did not change substantially, whereas the population's attitudes toward science teaching indicated substantial change after both phases of the course.

The anlayses that follow describe the differences in preservice elementary teacher attitudes toward science and science teaching. Pretest, posttest, and postposttest measures were taken according to the research design listed in Chapter III, Figure 1. Attitudes toward science and science teaching were measured by Richard Moore's (1973) Science Teaching Attitude Scales (STAS, Appendix E).

Science attitude findings. Scores on the STAS instrument can range from 0 to 120 in terms of attitudes toward science. The preservice elementary teachers' actual pretest, posttest, and postposttest attitudes toward science ranged from 66 to 91, 71 to 95, and 67 to 97, respectively (Table 1). The population means increased 2.76 from a pretest mean of 78.64 to a posttest mean of 81.40 and decreased -0.32 from the posttest to a postposttest mean of 81.08.

Changes between the means were not considered substantial. An examination of all individual preservice elementary teacher scores and science attitude changes revealed that more than 75% of the population experienced attitude score changes of greater than three points in total attitudes toward science. This change was interpreted as evidence that one could expect a majority of the

Table 1

Science Attitude Statistics

Statistics	Pretest	Posttest	Postposttest
Mean	78.64	81.40	81.08
Standard Deviation	7.40	6.23	8.25
Median	77.25	82.83	80.50
Maximum	91.00	95.00	97.00
Minimum	66.00	71.00	67.00
Range	25.00	24.00	30.00

population to experience attitude changes of at least three or more points and led to the construction of a ± 3 point referent to be used as a measure of substantial attitude change. Applying the referent to the changes in population means revealed that no substantial changes in the population's attitudes toward science occurred throughout the study. On this basis, these findings fail to support the rejection of Null Hypothesis 1.

Science teaching attitude findings. The range of the STAS instrument is 0 to 90 for total attitudes toward science teaching. The actual ranges for preservice elementary teacher pretest, posttest, and postposttest attitudes were 36 to 64, 44 to 71, and 52 to 79, respectively (Table 2). The population means increased 6.16 points from a pretest mean of 53.64 to a posttest mean of 59.80, and 4.40 from the posttest mean to a postposttest mean of 64.20.

Individual science teaching attitude scores and the degree of changes were compared. Like the science attitude scores, more than 75% of the population experienced attitude score changes of three or more points in total attitudes toward science teaching. Thus, a change of ± 3 points was considered a substantive change and was used as a referent. Applying the referent to the changes in population means revealed that substantive changes in the population's attitudes toward science teaching occurred throughout the study. These findings support the rejection of Null Hypothesis 1.

Population changes in attitudes toward science and science teaching were not equal. Substantial changes in attitudes toward science did not occur, whereas changes in attitudes toward science teaching were substantive. For a more detailed explanation of changes

Table 2

Science Teaching Attitude Statistics

Statistics	Pretest	Posttest	Postposttest
Mean	53.64	59.80	64.20
Standard Deviation	6.09	6.06	7.08
Median	53.00	59.25	65.50
Maximum	64.00	71.00	79.00
Minimum	36.00	44.00	52.00
Range	28.00	27.00	27.00

in attitudes toward science and science teaching, the reader is encouraged to examine the following additional findings before proceeding to Question 2 found on page 67.

Additional science attitude findings. Examination of changes only in population means does not permit one to examine additional important changes which occurred in attitudes. Therefore, the following findings are offered for a more complete explanation of changes which occurred in the population's attitudes toward science.

Science attitudes in the population were ranked as "high," "moderate," and "low" in terms of 15 point intervals on the STAS scales by using the following ranges:

High Science Attitude Scores = 105 - 120
Moderate Science Attitude Scores = 90 - 104
Low Science Attitude Scores = < 90

Further analysis revealed that none of the preservice teachers scored in the "high" science attitude range. From pretest to posttest to postposttest, the "moderate" scores increased 4%, 12%, and 20%, respectively. "Low" scores decreased from 96% to 88% to 80%.

The ± 3 point referent was used as evidence of substantive change and was utilized as a point of reference when percentages of population changes were computed and profiles of individual attitude changes were charted. From pretest to posttest, 60% of the population experienced an increase in science attitudes. Positive changes ranged from 3 to 15 points with a positive change mean of 7.23. Twenty-four percent of the population experienced a decrease in science attitudes, ranging from -3 to -13, with a negative change mean of -7.42. Sixteen percent of the population experienced no science attitude change.

From posttest to postposttest the percentage of positive changes was 28%, and ranged from 4 to 14 points with a positive change mean of 8.43. Forty-four percent of the population experienced negative attitude changes, ranging from -3 to -10.5, with a negative change mean of -6.23. The remaining 28% of the population experienced no change in science attitude.

Science attitude changes were analyzed further by examining individual profiles. Changes of three or more points were used to indicate positive, negative or neutral attitude changes. Positive science attitude changes were experienced by 60% of the population during the on-campus instruction and 28% during the field experience. The remainder of the population experienced negative attitude changes of 24% and 44% with 16% and 28% remaining unchanged.

Of the science attitudes measured after campus-instruction, 57% of the positive changes were of a moderate to high degree. Negative changes were evenly divided with 50% moderate to high and 50% low degree changes. Attitudes measured after the field experience revealed that 71% of the positive changes were moderate to high. Of the negative science attitude changes 54.5% were moderate to high and 45.5% were low degree attitude changes.

Five individuals displayed high to moderate degree positive science attitude changes after the field experience. Four of the five had initially experienced low to moderate negative attitude changes and one had experienced a low positive attitude change. Three of the six individuals displayed moderate to high negative attitude changes and previously experienced high positive changes, one had displayed a moderate positive change, and two experienced low positive changes in science attitudes.

Additional science teaching attitude findings. The following findings are offered for a more complete explanation of changes which occurred in the population's attitudes toward science teaching.

Science teaching attitudes in the population were ranged as "high," "moderate," and "low" in terms of 15 point intervals on the STAS by using the following ranges:

High Science Teaching Attitude Scores = 75 - 90
Moderate Science Teaching Attitude Scores = 60 - 74
Low Science Teaching Attitude Scores = < 60

Additional analysis revealed that none of the preservice teachers scored in the "high" science teaching attitude range on the pretest and posttest. However, 8% of the preservice teachers scored in the "high" attitude range on the postposttest. "Moderate" scores increased from 20% on the pretest to 48% and 64% on the posttest and postposttest. "Low" range scores were decreased from 80% to 52% to 28%, respectively. Attitude score changes are also examined as positive, negative, and neutral.

A change of ± 3 points was considered a substantial change and was used as a referent when computing percentages of population changes and developing individual attitude change profiles. A 76% positive change in attitudes toward science teaching was in evidence after the on-campus treatment (pretest and posttest). Nineteen individual positive changes resulted. They ranged from 3 to 21 points, with a change mean of 8.50. Three individauls (12%) experienced decreases in attitudes which ranged from -3 to -5. Twelve percent of the population's attitudes were unchanged.

From posttest to postposttest the percentages of positive attitude change was 56% and ranged from 4 to 22 points. The positive change mean was 9.86. The negative change percentage remained the same, 12% and ranged from -4 to -7. Thirty-two percent of the population's attitudes were unchanged.

Of the attitudes toward science teaching measured after campus methods instruction 84.2% of the positive attitude changes were of a moderate to high degree. All of the negative attitude changes were low degree, accounting for 12% of the population.

Attitudes measured after the field experience revealed that 79% of the positive attitude changes were moderate to high, compared with 67% moderate and 33% low degree negative attitude changes.

Fifteen individuals displayed moderate to high degree attitude changes after the posttest. After the postposttest, six of the fifteen individuals experienced no attitude changes. Two, four, and one individual displayed high, moderate, and low changes, respectively. Two individuals exhibited moderate negative changes and one experienced a low negative attitude change.

Question 2

Which communicator was perceived as being most credible during and following the methods course?

The order of ranks for communicators in terms of perceptions from most to least credible were: science methods instructor, university supervisor, graduate assistant, unit evaluator, peer team member and cooperating teacher. The credibility ranks did not change from posttest to postposttest even though the means of all percentages of communicator credibility increased.

Specific procedures were used to identify the most credible communicator during and following the methods course. Since the preservice teachers were not familiar with the communicators at the beginning of the quarter, no pretest of communicator credibility was given. Reports of changes in credibility perceptions and communicator ranks, from most to least credible, were based on posttest and post-posttest data.

Posttest credibility findings. The Perception of Communicator Credibility instrument, a semantic differential, (Appendix G) provided individual ratings for six communicators: unit evaluator, graduate assistant, university supervisor, cooperating teacher, science instructor, and peer team member. Each communicator was rated on the concepts: (1) expertness and (2) trustworthiness in science and science teaching. Ratings were marked on a seven space semantic differential with scores that ranged from -3 to +3 on each of the nine scales. The middle space of the distance was considered neutral. Ratings for each of the nine bipolar adjectives were averaged for each concept. Because Hovland reported that expertness and trustworthiness were inseparable components of credibility, the researcher made no attempt to report separate scores for each of the two concepts. Scores for the two concepts were averaged and considered an individual's perception of a communicator's credibility.

Results for the posttest given after the campus instruction during week seven of the quarter are reported in Table 3. During this period, the preservice elementary teachers were in contact with each of the six communicators, but spent more time with the science instructor, graduate assistant, and unit evaluators. Rankings of

Table 3

Descriptive Statistics of Posttest Perceptions
of Communicator Credibility

Statistics	Science Instructor	University Supervisor	Graduate Assistant	Unit Evaluator	Peer Team Member	Cooperating Teacher
Sum of Scores	65.93	61.71	58.00	47.03	44.14	25.78
N	25	25	25	25	24	25
Mean	2.64	2.47	2.32	1.88	1.84	1.03
Standard Deviation	0.44	0.61	0.75	1.03	0.91	0.77
Median	2.89	2.67	2.50	2.28	1.78	0.91
Maximum	3.00	3.00	3.00	3.00	3.00	2.56
Minimum	1.62	1.28	0.45	0.06	-0.61	-0.39
Range	1.38	1.72	2.55	2.94	3.61	2.95

the communicators from most to least credible are presented in
Table 4. The rankings are based on the population means previously
reported in Table 3 and support the rejection of Null Hypothesis 2.

Postposttest credibility findings. Results are shown in
Table 5. The postposttest was given during week eleven of the
quarter, four weeks after the posttest. During those four weeks,
the preservice teachers were in the field and in close contact with
the following communicators: university supervisor, cooperating
teacher, and peer team member. Communication with the other communicators was minimal or did not exist.

The population's credibility rankings for the six communicators
are identified in Table 6. Rankings are based on the population means
reported in Table 4. The population rankings remained the same from
posttest to the postposttest despite increases in credibility for all
communicators. Changes in the sums and means of credibility scores
are reported in Table 7 and support the rejection of Null Hypothesis 2.

The order of ranks based upon preservice elementary teacher
perceptions of communicator credibility did not change from posttest
to postposttest. The ranks were: (1) science methods instructor,
(2) university supervisor, (3) graduate assistant, (4) unit evaluator,
(5) peer team member, and (6) cooperating teacher. For a more detailed
study of communicator credibility, the reader is encouraged to refer
to supplementary tabular data reported in the additional findings
which follow before proceeding to Question 3 found on page 74.

Additional posttest credibility findings. Additional analysis
revealed that many individuals rated several communicators equally high
in credibility. Despite this the population rankings did not vary with

Table 4

Posttest Population Perceptions of Communicator Credibility Rank

Credibility	Rank	Communicator
Most	1	Science Instructor
↑	2	University Supervisor
	3	Graduate Assistant
	4	Unit Evaluator
↓	5	Peer Team Member
Least	6	Cooperating Teacher

Table 5

Descriptive Statistics of Postposttest Perceptions
of Communicator Credibility

Statistics	Science Instructor	University Supervisor	Graduate Assistant	Unit Evaluator	Peer Team Member	Cooperating Teacher
Sum of Scores	67.48	64.87	63.16	54.54	45.16	42.28
N	25	25	25	25	24	25
Mean	2.70	2.59	2.53	2.18	1.88	1.69
Standard Deviation	0.41	0.48	0.59	0.83	1.04	1.06
Median	2.98	2.84	2.85	2.33	2.00	2.00
Maximum	3.00	3.00	3.00	3.00	3.00	3.00
Minimum	1.73	1.34	1.00	0.22	-1.06	-0.61
Range	1.27	1.66	2.00	2.78	4.06	3.61

Table 6

Postposttest Population Perceptions of Communicator Credibility Rank

Credibility	Rank	Communicator
Most	1	Science Instructor
↑	2	University Supervisor
	3	Graduate Assistant
	4	Unit Evaluator
	5	Peer Team Member
Least	6	Cooperating Teacher

Table 7

Posttest-Postposttest Changes in Communicator Credibility

	Posttest		Postposttest		Difference	
Communicator	Sum	Mean	Sum	Mean	Sum	Mean
Science Instructor	65.93	2.64	67.48	2.70	1.55	0.06
University Supervisor	61.71	2.41	64.87	2.59	3.16	0.12
Graduate Assistant	58.00	2.32	63.16	2.53	5.16	0.21
Unit Evaluator	47.03	1.88	54.54	2.18	7.51	0.30
Peer Team Member	44.14	1.84	45.16	1.88	1.02	0.04
Cooperating Teacher	25.78	1.03	32.28	1.69	16.50	0.66

those reported in Table 4. Individual rankings of communicator credibility are important in the analysis of Null Hypothesis 3.

Additional postposttest findings. As measured by the postposttest, many individuals perceived the science and science teaching credibilities of several different communicators to be equal. Even so, the population rankings did not vary from those reported by Table 6, nor did they vary from the posttest rankings reported by Table 4. The individual communicator credibility rankings were used to assist the computation of perceived communicator science and science teaching attitudes for testing Null Hypothesis 3.

Question 3

Did the preservice elementary teachers' attitudes change toward and approach the perceived attitude level of the most credible communicator (PCA)?

An examination of attitude change profiles revealed that a majority of the population's attitudes toward science and science teaching changed toward and approached the PCA. However, 32% of the population approached the PCA early in the course and later experienced decreases in attitudes toward science.

The following analyses compare the preservice teachers' science and science teaching attitude scores to their perceptions of the most credible communicators' attitudes (PCA). Preservice teacher attitude scores were measured by the STAS instrument at pretest, posttest, and postposttest times.

Preservice teacher preceptions of communicator science and science teaching attitudes were measured by the STAS and reported on the Perceptions of Communicator Attitude (PCA) answer forms

(Appendix F). Perceptions of science instructor and graduate assistant attitudes were measured during week seven of the quarter after the methods instruction. Perceptions of the remaining communicators' attitudes were measured during week eleven shortly after the completion of the field experience.

Findings of perceived communicator attitudes toward science. Pretest, posttest, and postposttest individual attitude scores were compared with the perceived score of the most credible communicator(s). Profiles of the direction and degree of attitude changes were developed and appear in Table 8.

Profile 1 indicates that 40% of the population's science attitudes changed in the direction of and approached the most credible communicator's perceived level of attitude. More than half of the individuals identified by Profile 1 exceeded the perceived level of communicator attitude (PCA).

According to Profile 2, 16% of the population did not experience attitude changes in the direction of or attain the PCA level of the most credible communicator. Profile 3 indicates that 32% of the population experienced pretest to posttest attitude direction changes toward the PCA level. Increased science attitudes were, however, followed by decreases from posttest to postposttest. The converse is illustrated by Profile 4. Twelve percent of the preservice elementary teachers' attitude directions and levels initially decreased, but from posttest to the postposttest the attitudes moved toward and attained the PCA. Particularly on the basis of Profiles 1 and 3, these findings in general support the rejection of Null Hypothesis 3.

Table 8

Science Attitude Changes and Perceptions of Communicator Attitudes

Profile Number	Profiles*				Frequency	Frequency Percent	Identification Numbers
	PRE	POST	PPOST	PCA			
1					10	40	1*, 2, 4*, 5*, 8*, 13*, 14, 16, 21, 25*
2					4	16	3, 15, 18, 22
3					8	32	6, 7, 9, 10, 11, 12, 17, 24
4					3	12	19, 20, 23

*Attitudes exceeded perceived attitudes of most credible communicator.

Findings of perceived communicator attitudes toward science teaching. Profiles of science teaching attitude change direction and level are available in Table 9.

Profile 1 shows that 72% of the population's science teaching attitudes changed in the direction and approached the perceived attitude levels of the most credible communicator. Twelve percent of the preservice teachers' science teaching attitudes exceeded the PCA.

As illustrated by Profile 2, 8% of the population did not experience attitude changes toward or approach the level of the most credible communicator. The science teaching attitudes of these two individuals remained relatively stable throughout the treatments.

Profiles 3 and 4 are similar to those found in Table 8. According to Profile 3, 12% of the population's science teaching attitudes moved toward and attained the perceived attitude level from pretest to posttest, but decreased from posttest to postposttest. The opposite occurred for those individuals represented by Profile 4. Initial pretest to posttest decreases were accompanied by posttest to postposttest movement toward and attainment of the perceived communicator attitude level by 8% of the population. Collectively these findings support the rejection of Null Hypothesis 3.

Question 4

Were the factors of content area specialization groups, self-perceptions of science training, and achievement and number of completed science courses related to the preservice elementary teachers' attitudes toward science and science teaching?

Perception of science training was the only factor found related to attitudes toward science and science teaching. All of the remaining

Table 9

Science Teaching Attitude Changes and Perceptions of
Communicator Attitudes

Profile Number	Profiles*				Frequency	Frequency Percent	Identification Numbers
	PRE	POST	PPOST	PCA			
1					18	72	1, 2*, 3, 4, 5, 6, 7*, 8, 9, 11, 12, 14, 15, 16, 17, 18, 21, 24*
2					2	8	20, 23
3					3	12	10, 22, 25
4					2	8	13, 19

*Attitudes exceeded perceived attitudes of most credible communicator.

respondent factors were related to attitudes toward science teaching. The analyses that follow describe the relationships between preservice elementary teacher attitudes and the perceived attitudes of the most credible communicator (PCA) relevant to the areas of program specialization, perceptions of science training, and science achievement.

Specialization groups. Individuals in each of the six elementary education program specialization groups were identified: language arts, social studies, mathematics, early childhood, special education, and science. Individual sicence and science teaching attitude scores, the perceived attitude scores of the most credible communicator (PCA), and the group means for each specialization area were statistically analyzed. Differences between the test means and the PCA for each group are found in Table 10 for attitudes toward science and Table 11 for attitudes toward science teaching.

All groups demonstrated positive changes in science attitudes ranging from 0.40 to 11.50 points. Increases in science teaching attitudes ranged from 0.50 to 16.33. The social studies group demostrated the greatest growth on both attitude measures. With the exception of the science group, all groups experienced increases in science teaching attitudes (see Tables 10 and 11).

Tables 12 and 13 illustrate the attitude changes of each specialization group. The attitude change profiles of Table 12 reflect dissimilarity among the groups' attitude changes toward the perceived attitude level of the most credible communicator. The profile of the social studies group represents a consistent attitude change throughout the treatments and matches the PCA better than all groups, except special education. Special education and science group profiles

Table 10

Comparison of Specialization Group and PCA Science Attitudes

lization oup	PRE Mean	POST Mean	Difference	PPOST Mean	Difference	PPOST-PRE Difference	PCA Mean
ge Arts	82.20	79.10	-3.10	84.50	5.40	2.30	89.20
Studies	76.67	81.00	4.33	88.17	7.17	11.50	87.50
atics	76.00	84.33	8.33	78.67	-5.66	2.67	82.00
Childhood	77.71	84.21	6.50	78.36	-5.85	0.65	85.14
l Education	81.40	81.50	0.10	81.80	0.30	0.40	81.96
e	84.00	82.00	-2.00	84.50	2.50	0.50	80.00

Table 11

Comparison of Specialization Group and PCA Science Teaching Attitudes

lization oup	PRE Mean	POST Mean	Difference	PPOST Mean	Difference	PPOST-PRE Difference	PCA Mean
ge Arts	56.80	61.80	5.00	65.90	4.10	9.10	74.20
Studies	51.67	59.33	7.66	68.00	8.67	16.33	73.67
atics	53.00	59.67	6.67	64.00	4.33	11.00	65.00
Childhood	51.00	62.00	11.00	61.93	-0.07	10.93	70.61
l Education	53.80	56.20	2.40	65.40	9.20	11.60	65.00
e	61.00	55.50	-5.50	61.50	6.00	0.50	66.75

Table 12

Specialization Group Science Attitude Profiles

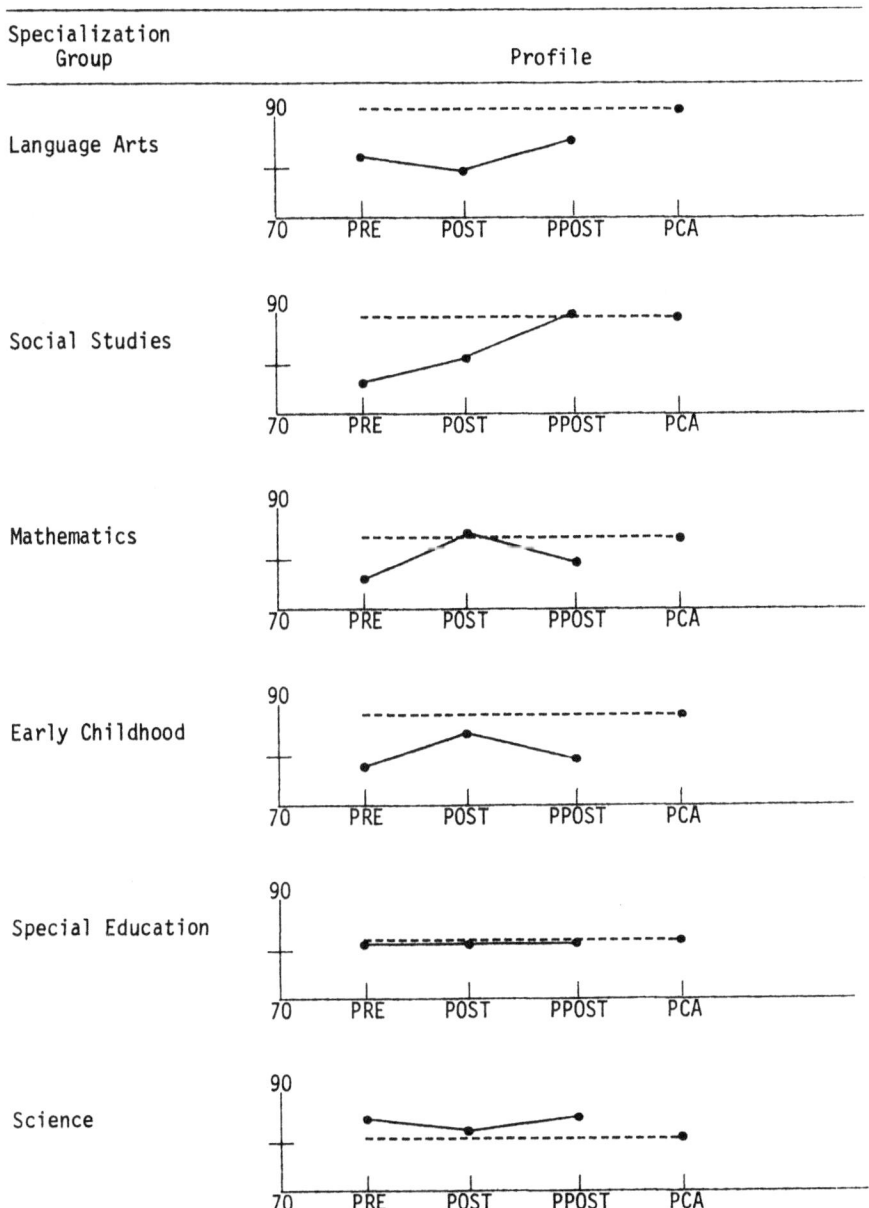

Table 13

Specialization Group Science Teaching Attitude Profiles

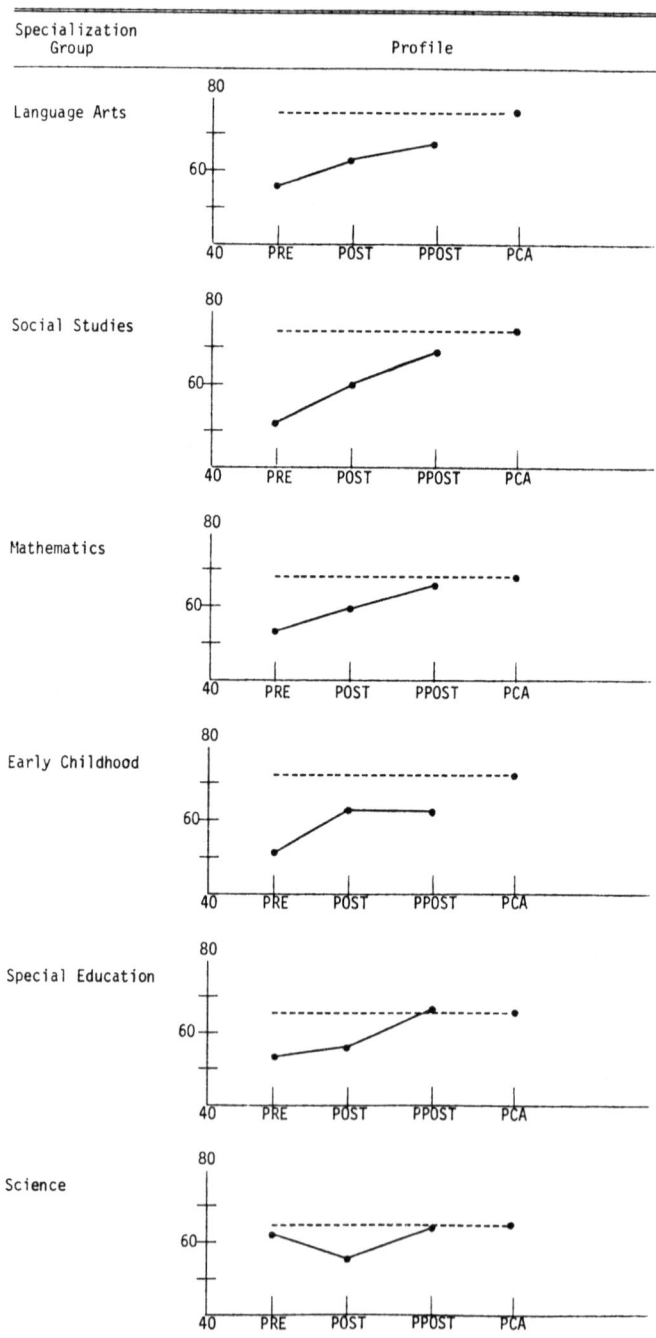

reflect attitudes which remained virtually unchanged throughout the treatment. The science attitude profile of the other groups were more varied.

Group attitudes toward science teaching, as profiled in Table 13, varied less and reflected linear attitude changes toward the PCA than the science attitude profiles of Table 12. The differences between the means of the groups' science teaching attitudes and the mean PCA (Table 11) were greater than the differences between group science attitudes and group PCA's (Table 10). The above findings fail to support rejection of Null Hypothesis 4 for changes in attitudes toward science, but support the rejection of the null hypothesis for attitude changes toward science teaching.

Perceptions of science training. Individual perceptions of science training were obtained from the 5-P and 5-N scales of the STAS (Chapter III, page 47). The range of each scale is 0-15. When combined these scales have a raw score range of 0-30 for this measure. The population's perceptions improved from a mean of 18.64 to 21.92 to 24.76 on the combined scales (Table 14).

The postposttest distribution of scores was used to divide the population into four perception of science training groups: "high," "above average," "average," and "low." Group means were examined and differences in science and science teaching attitudes are reported in Tables 15 and 16.

With the exception of the "average" perception group, all groups demonstrated positive changes in science attitudes. The "high" and "above average" groups achieved the most substantial increases. Science attitude change profiles are found in Table 17 and illustrate

Table 14

Population Changes in Perception of Science Training From
Pretest to Posttest to Postposttest

Statistics	Pretest	Posttest	Change	Postposttest	Change	Total Change
Mean	18.64	21.92	3.28	24.76	2.84	6.12
Standard Deviation	4.72	4.81	0.15	4.73	-0.14	0.01
Median	19.67	21.00	1.33	26.00	5.00	6.33
Maximum	27.00	30.00	3.00	30.00	0.00	3.00
Minimum	6.00	13.00	7.00	14.00	1.00	8.00
Range	21.00	17.00	-4.00	16.00	-1.00	-5.00

the relationship between group changes, direction, and the degree to which attitudes approached the PCA. "High" and "above average" groups demonstrated similar profiles in terms of direction and degree of attitude change. The "average" group's profile indicates a slight decrease in attitude and no movement toward the PCA; whereas the profile of the low perception group indicates an initial increase in excess of the PCA that is followed by a post to postposttest science attitude decrease.

All groups experienced increases in science teaching attitudes ranging from 4.60 to 12.33 points. The "high" and "above average" perception groups demonstrated the most attitude changes, 12.33 and 11.73 points respectively (Table 16). All but the "low" perception group maintained profiles that moved toward the level of the PCA (Table 18). Movement toward the PCA was related to the degree of science training perception. The "low" perception group's attitude profile was varied with changes of 9.66 and -2.00 points from pre to post to postposttest, and displayed the least movement of all groups toward the PCA. Collectively, these findings support the rejection of Null Hypothesis 4 for attitude changes toward science and science teaching.

Science achievement. The number of science courses completed and the grades received in each were obtained from student records. All grades received for science course work were averaged into a grade point average (GPA) for each individual. The population's grades ranged from 4.00 to 1.00 in science achievement with a GPA mean of 2.73 and standard deviation of 0.78. The number of science courses completed ranged from 4 to 1 with a mean of 2.80 and standard deviation of 0.94.

Table 15

Comparison of Science Training Perception Groups and Science Attitudes

Perception Group	PRE Mean	POST Mean	Difference	PPOST Mean	Difference	PPOST-PRE Difference	PCA Mean
High	83.33	86.50	3.17	87.92	1.42	4.59	88.17
Above Average	76.09	78.64	2.55	79.14	0.50	3.05	82.33
Average	79.40	79.00	-0.40	78.60	-0.40	-0.80	85.20
Low	77.33	82.00	4.70	78.67	-3.33	1.34	81.00

Table 16

Comparison of Science Training Perception Groups and Science Teaching Attitudes

Perception Group	PRE Mean	POST Mean	Difference	PPOST Mean	Difference	PPOST-PRE Difference	PCA Mean
High	56.50	65.67	9.17	68.83	3.16	12.33	66.8
Above Average	52.91	57.55	4.64	64.64	7.09	11.73	68.21
Average	55.40	58.40	3.00	60.00	1.60	4.60	70.90
Low	47.67	57.33	9.66	55.33	-2.00	7.66	75.67

Table 17

Science Training Perception Group Science Attitude Profiles

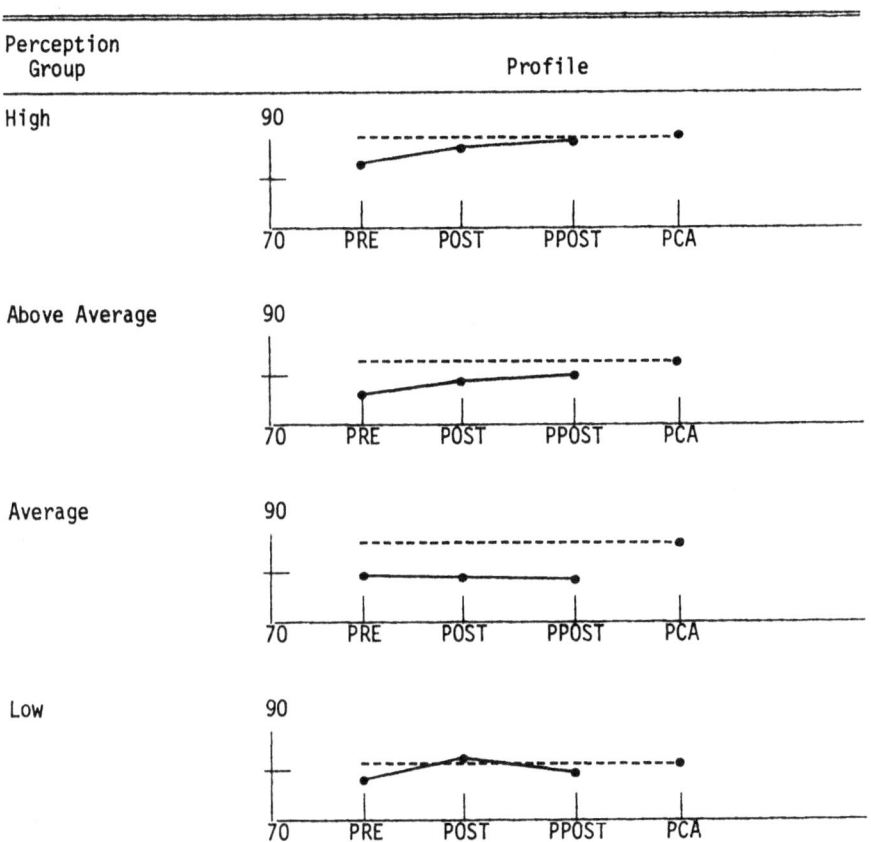

Table 18

Science Training Perception Group Science Teaching Attitude Profiles

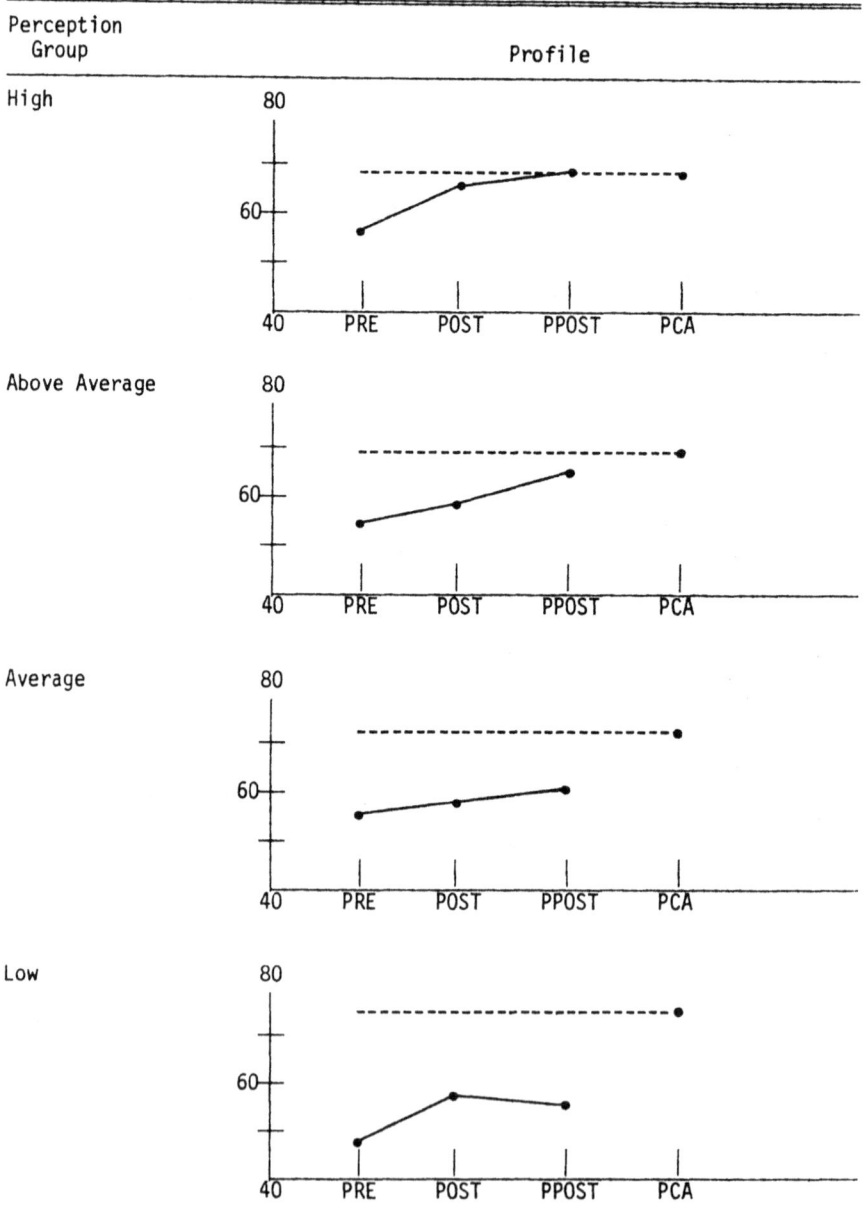

The preservice teachers were divided into "high," "above average," "average," and "low" ability groups by an examination of the population frequency distribution. Sixteen percent of the population ranked in the "high" ability group, 40% were "above average," 36% "average," and 8% ranked "low." Group differences between test means and the PCA are contained in Table 19 for science attitudes and Table 20 for science teaching attitudes. Group science attitude profiles are illustrated in Table 21. All group profiles vary and none represent science attitude changes toward the level of the PCA.

The science teaching attitude profiles of Table 22 are similar in each group. All achievement group profiles moved toward the PCA, but the "high," "above average," and "average" groups more closely approached the PCA level. The above findings fail to support rejection of Null Hypothesis 4 for changes in attitudes toward science, but support rejection of the null hypothesis for attitude changes toward science teaching.

Number of science courses completed. The preservice elementary teachers also were divided into groups based upon the number of science courses completed before the treatments. Science, science teaching attitudes, and perceptions of communicator attitudes were examined for relationships according to group means and attitude changes. Comparisons of group means from pre to post to postposttest and comparisons with the group PCA are shown in Table 23 for science attitudes and Table 24 for science teaching attitudes.

Group science attitude change profiles are found in Table 25. Members of the groups who completed one and four science courses

Table 19

Comparison of Achievement Group and PCA Science Attitudes

Achievement Group	PRE Mean	POST Mean	Difference	PPOST Mean	Difference	PPOST-PRE Difference	PCA Mean
High	84.00	81.50	-2.50	82.25	0.75	-1.75	83.08
Above Average	77.30	82.80	5.20	79.55	-3.50	2.25	85.38
Average	75.56	72.44	-3.12	81.94	9.50	6.38	82.98
Low	79.50	79.00	-0.50	78.50	-0.50	-1.00	82.00

Table 20

Comparison of Achievement Group and PCA Science Teaching Attitudes

Achievement Group	PRE Mean	POST Mean	Difference	PPOST Mean	Difference	PPOST-PRE Difference	PCA Mean
High	56.50	58.50	2.00	64.00	5.50	7.50	71.50
Above Average	53.70	62.00	8.30	67.45	5.45	13.75	68.26
Average	53.67	58.89	5.22	63.66	41.7	9.39	68.28
Low	47.50	55.50	8.00	56.00	0.50	8.50	75.00

Table 21

Science Achievement Group Science Attitude Profiles

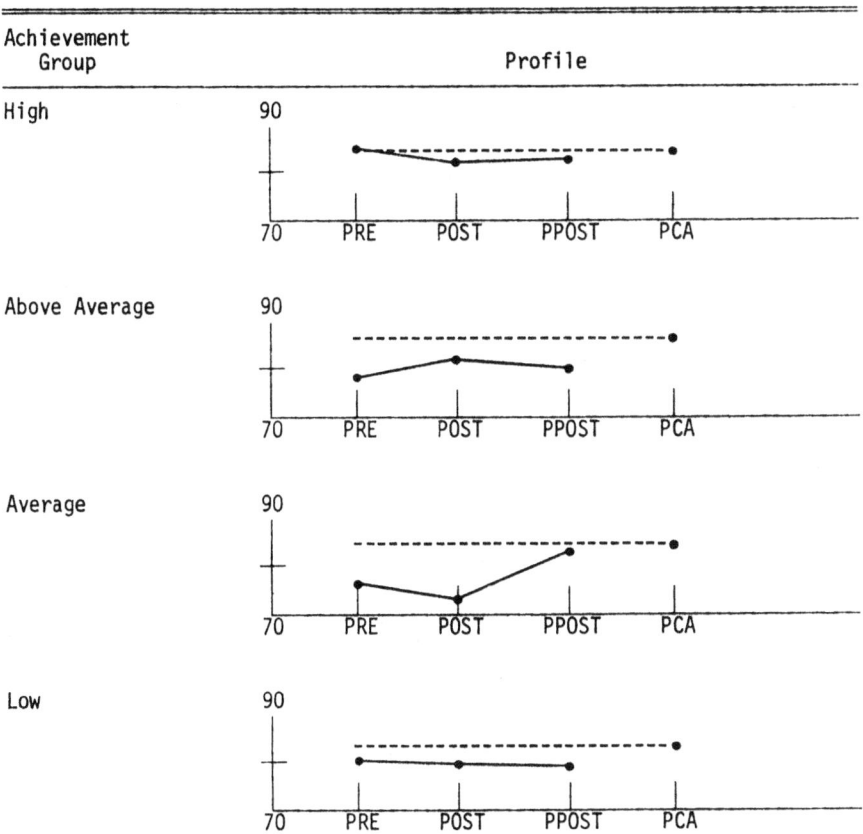

Table 22

Science Achievement Group Science Teaching Attitude Profiles

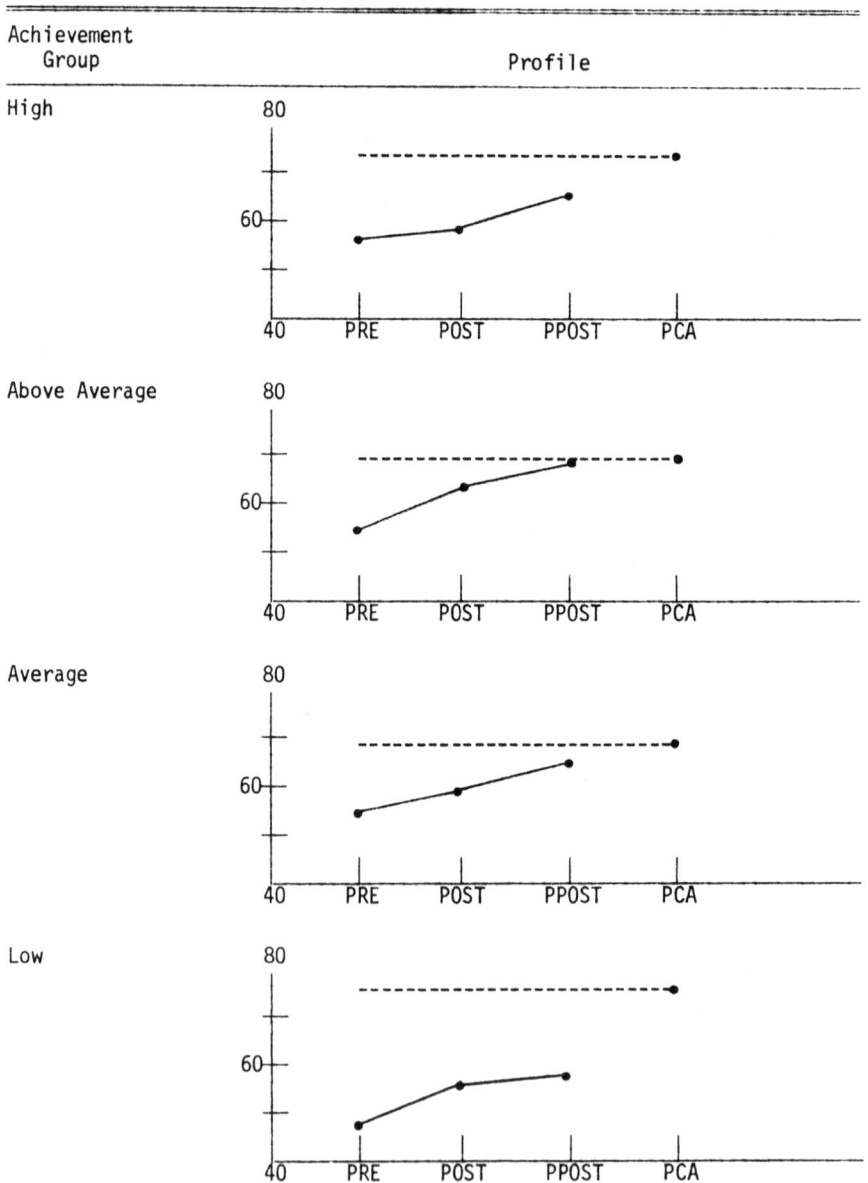

demonstrated attitude changes toward and attained the PCA level.
Individuals of the groups who completed two and three science courses
experienced initial attitude increases, then demonstrated decreased
attitudes from the post to postposttest.

Group science teaching attitude change profiles are illustrated in Table 26. All groups demonstrated increased attitudes toward science teaching throughout the treatments. Members in the group who had completed one science course exceeded the level of the PCA. The remaining groups approached the PCA level in proportion to the number of science courses completed. The above findings fail to support the rejection of Null Hypothesis 4 for changes in attitudes toward science, but support the rejection of the null hypothesis for attitude changes toward science teaching.

Summary

A summary of the relevant findings for each question presented in Chapter IV follows.

Question 1. Did the preservice elementary teachers' attitudes toward science and science teaching change during and following the treatments?

Attitudes toward science did not change substantially, whereas the population's attitudes toward science teaching indicated substantial change after all treatments.

A ± 3 point referent of substantial change was used as an indicator. The attitudes toward science population mean increased 2.76 from pretest to posttest and decreased -0.32 from posttest to postposttest. The population mean for attitudes toward science teaching increased 6.16 from pretest to posttest and 4.40 from posttest to postposttest.

Table 23

Comparison of Completed Science Course Groups and PCA Science Attitudes

of Group	PRE Mean	POST Mean	Difference	PPOST Mean	Difference	PPOST-PRE Difference	PCA Mean
	83.71	82.16	-1.55	85.86	3.70	2.15	85.07
	76.44	81.89	5.45	79.72	-2.17	3.28	85.00
	76.71	80.00	3.29	77.85	-2.15	1.14	82.87
	77.50	79.00	1.50	81.75	2.75	4.25	81.50

Table 24

Comparison of Completed Science Course Groups and PCA Science Teaching Attitudes

of Group	PRE Mean	POST Mean	Difference	PPOST Mean	Difference	PPOST-PRE Difference	PCA Mean
	55.86	62.43	6.57	66.43	4.00	10.57	68.29
	53.89	61.33	7.44	62.67	1.34	8.78	68.26
	50.43	56.43	6.00	63.29	6.86	12.86	72.47
	56.00	60.50	4.50	69.00	8.50	13.00	67.88

Table 25

Completed Science Course Group Science Attitude Profiles

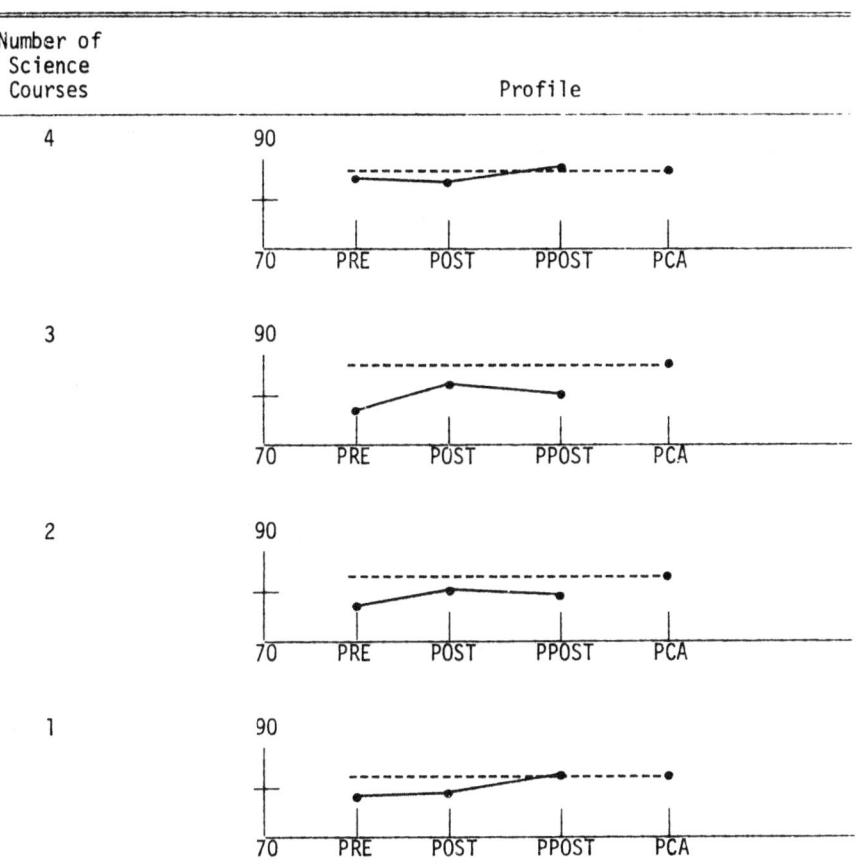

Table 26

Completed Science Course Group Science Teaching Attitude Profiles

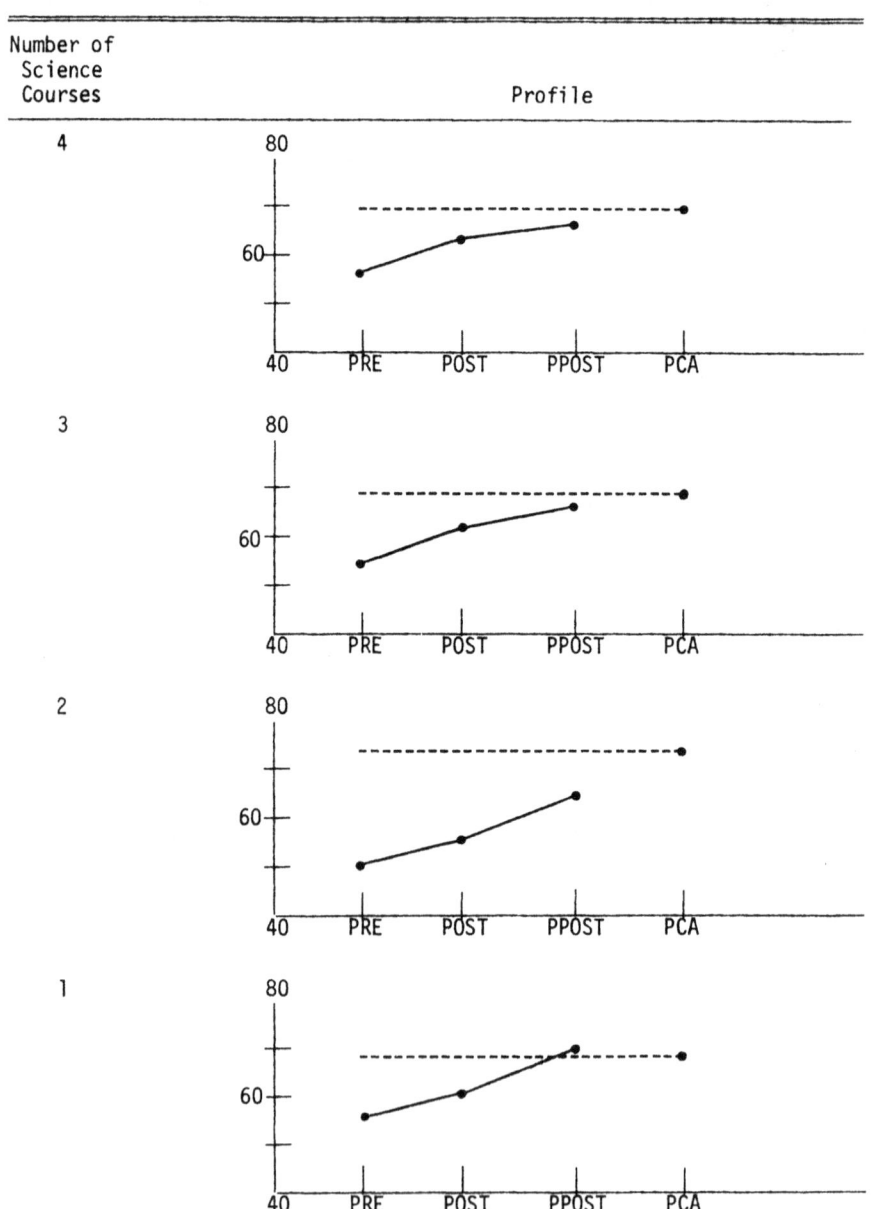

Question 2. Which communicator was perceived as being most credible during and following the methods course?

The rankings of communicators in terms of perceptions from most to least credible were: science methods instructor, university supervisor, graduate assistant, unit evaluator, peer team member, and cooperating teacher. The credibility ranks did not change from posttest to postposttest even though the means of all perceptions of communicator credibility increased. Cooperating teachers displayed the greatest change in credibility, 16.50, and peer team members the least, 1.02. Communicators and frequencies of those rated most credible on the postposttest included: science methods instructor, university supervisor, and graduate assistant.

Question 3. Did the preservice elementary teachers' attitudes change toward and approach the perceived attitude level of the most credible communicator (PCA)?

An examination of attitude change profiles revealed that, in general, the population's attitudes toward science changed toward and approached the PCA, whereas a majority of the population's attitudes toward science teaching more clearly changed toward and approached the PCA level.

Comparisons of individual attitude change profiles indicated that more (72%) preservice teachers experienced science teaching attitude changes toward the PCA than changes in attitudes toward science (40%). Sizeable portions of the population experienced science and science teaching attitude change movements toward the PCA from pretest to posttest (32% - science; 12% - science teaching or posttest to postposttest (12% - science; 8% - science teaching).

Question 4. Were the factors of content area specialization groups, self-perceptions of science training, and achievement and number of completed science courses related to the preservice elementary teachers' attitudes toward science teaching?

Perception of science training was the only factor found related to attitudes toward science and science teaching. All remaining factors were related to attitudes toward science teaching.

Relationships were determined by an examinaton of profiles generated by group mean attitude score movements toward perceptions of communicator attitudes (PCA). Specialization group attitude changes toward the PCA were mixed for science attitudes, while four of six groups experienced science teaching attitude changes toward the PCA: language arts, social studies, mathematics, and special education.

The population's perception of science training mean increased on all measures for a total of 6.12 raw score points on a scale which ranged from 0 to 30. Group profiles revealed that "high" and "above average" perception groups moved toward the PCA for both science and science teaching attitudes, whereas "average" and "low" perception groups displayed limited movements.

Group profiles for grade point averages (GPA) and completed science courses were mixed with no strong relationship to science attitude change and PCA. All GPA groups demonstrated science teaching attitude changes toward the PCA, with all groups but low GPA approaching the PCA level. Groups based upon the number of completed science courses maintained profiles similar to those described for GPA groups. Profiles of changes in science attitudes were varied, whereas all group science teaching attitude change profiles approached the PCA.

Answers and findings for each research question were provided in Chapter IV. Chapter V consists of three sections: Conclusions,

Implications, and Recommendations. All conclusions are supported
by an appropriate rationale and summarized for use in implications
of findings and recommendations for future research.

Chapter V

CONCLUSIONS, IMPLICATIONS, AND RECOMMENDATIONS

Overview

Chapter V is composed of three sections: Conclusions, Implications, and Recommendations. The first section summarizes the findings by answering each question posited in Chapter III and provides a rationale for each finding. The findings are related to teacher education programs in the Implications section. The final section, Recommendations, offers suggestions for further research relevant to communicator credibility and preservice teacher attitude change.

Conclusions

<u>Question 1</u>: Did preservice elementary teachers' attitudes toward science and science teaching change during and following the team taught, competency-based, field oriented science methods course?

<u>Attitudes toward science</u>. As measured after the on-campus methods instruction (pretest to posttest), 60% of the population's attitudes toward science improved, 25% decreased, and 16% experienced no change. After the field experience (posttest to postposttest) 28% of the population's science attitudes improved, 44% declined, and 28% remained the same. The individual profiles of Table 32-I provided graphic interpretations of the science attitude changes which occurred throughout the study.

On the basis of these findings a majority of the population's science attitudes changed in a positive direction from pretest to posttest only. However, if the ± 3 point referent of significant change which was used to construct population attitude change profiles is applied to the measures, the population's science attitudes did not change appreciably during and following the methods course. The pretest to posttest population mean increased by 2.76 points and decreased 0.32 from posttest to postposttest (Table 1). On this basis, these findings fail to support rejection of Null Hypothesis 1. Therefore, it is concluded that the preservice elementary teachers' attitudes toward science did not change during and following the team taught, competency-based, field oriented science methods course.

Attitudes toward science teaching. After the methods instruction (pretest to posttest), 76% of the population experienced positive changes in science teaching attitudes, 12% displayed decreased attitudes and 12% remained unchanged. The population science teaching attitude mean increased 6.16 from a pretest mean of 53.64 to a posttest mean of 59.80 (Table 2). Measures obtained after the field experience (postposttest) revealed a 4.40 increase in attitudes (population mean of 64.20) for a total increase of 10.56 points on the science teaching attitude scales. From the posttest to postposttest measures, 50% of the population displayed positive attitude changes, 13% negative changes and 32% were unchanged. These findings, plus using the ± 3 point change referent, serve to support rejection of the null hypothesis and make possible the conclusion that the preservice elementary teachers' attitudes toward science teaching changed during and following the methods course.

Discussion. Several considerations are offered to explain the science and science teaching attitude change disparity and the larger percentages of attitude increases on both variables after the on-campus methods treatment. The three most credible communicators (i.e., science instructor, graduate assistant, and one other communicator who served in the dual role of unit evaluator and supervisor) closely coordinated their instruction and articulated very similar communications during the on-campus treatment. During this phase of the treatment, neither communicator spent lengthy periods of time espousing the virtues of science and discussing particular philosophies or theories of science. Because of the intent of the course, the communicators spent considerable time on philosophies and processes of science education relevant to contemporary elementary science curricula, the importance and role of science in the elementary curriculum, its relation to other curricular areas, science activities, science teaching methods, the incorporation of generic methods, classroom management, and effective science unit planning.

The validity of the science attitude scales could be suspect and may have contributed to the disparity in attitudes due to changes in contemporary philosophies of science. However, it is likely that the most credible communicators presented communications that more often addressed the attitudes which were measured by the science teaching attitude scales and less often addressed those items measured by the science attitude scale. For example, two of the three most credible communicators did not agree with the 3 P scale of the STAS and tended to rate positively the 3 N scale items when completing the STAS. It is suspected that these attitudes were transferred to the preservice teachers and may have served to depress the science attitude scores on these scales. Because the most credible communicators were educators first and scientists second, similar disagreements with the remaining science attitude scales may have influenced the population's attitude scores.

A further consideration resulted during the field experience phase of the treatment. The three most credible communicators did not have the opportunity for continued coordination and articulation of their messages. Also, a common forum did not exist wherein all preservice teachers could receive the same kinds of messages from the communicators or their special factor groups. Thus, this closely knit attitude support system broke down with little influence of attitudes the net result.

Question 2: Which communicator was perceived as being most credible during and following the methods course?

The population's perceptions of credibility rank did not change from posttest to postposttest. Thus, the communicators were ranked from most to least credible as follows:

1. Science Instructor
2. University Supervisor
3. Graduate Assistant
4. Unit Evaluator
5. Peer Team Member
6. Cooperating Teacher

These findings support rejection of Null Hypothesis 2. Yet, the science instructor cannot be claimed the most credible communicator for all members of the population. The graduate assistant and a university supervisor also received very high credibility ratings. The answer to Question 2 is not one particular communicator, but rather that of three principal communicators: Science Methods Instructor, Graduate Assistant and a Supervisor/Unit Evaluator. On the postposttest credibility measures, 68% of the population perceived more than one of the three principal communicators as being of equal credibility. For the purpose of answering Question 2, the Science Instructor can be claimed most credible. However, in this team teaching situation credibility recognition was often shared and should be considered in the study of attitude changes.

Question 3: Did the preservice elementary teachers' attitudes change in a direction and approach the perceived attitude level of the most credible communicator (PCA)?

Attitudes toward science. Profiles representative of individual attitude direction and level of changes were provided in Table 8. Forty percent of the population's attitudes changed toward, matched or exceeded the PCA. Thirty-two percent changed toward the PCA during the pretest-posttest interval, but did not maintain the attitude level during the posttest-postposttest interval, 12% changed toward the PCA during the posttest-postposttest interval, and 16% did not change toward the PCA throughout the treatments.

The research question and hypothesis did not specify when movement toward the PCA should occur. Considering Profiles 1 and 3 of Table 8, 72% of the population exhibited science attitude changes toward and approached the PCA level after one or both treatments. These findings represent positive relationships between individual attitudes and the science attitude perceptions of the most credible communicators. The existence of individual and PCA relationships supports rejection of Null Hypothesis 3. In general, these relationships provide an affirmative answer to Question 3; preservice elementary teacher attitudes changed in the direction and approaches the perceived attitude level of the most credible communicator at various times throughout the course.

Attitudes toward science teaching. The profiles of Table 9 provided individual science teaching attitude changes. Profile 1 is representative of the attitude changes experienced by 72% of the population -- steady increases of science teaching attitudes toward the PCA after both treatments. Profiles 3 and 4 revealed that 20% of the population experienced attitude changes toward the PCA after one treatment, but not the other. Only 8% of the individuals did not display science teaching attitude changes after the treatments. These findings represent relationships between individual science teaching attitude changes and the PCA at various times throughout the course. Existence of these relationships supports the rejection of the null hypothesis and the conclusion that preservice elementary teacher attitudes moved toward the perceived science teaching attitude level of the most credible communicator.

Question 4: Were the factors: specialization groups, self-perceptions of science teaching, and achievement in previous science courses related to the preservice elementary teachers' attitudes toward science and science teaching?

Specialization groups. Groups were formed from elementary content specialization areas which consisted of: language arts (5 individuals), social studies (3), mathematics (3), early childhood (7), special education (5), and science (2). Changes in the group attitude means indicated increased science and science teaching attitudes (except in the science group). The profiles of Tables 12 and 13 illustrated relationships between group attitude changes and PCA. However, one must be cautioned that some profiles may have been unduly influenced by small group size.

The profiles of Table 12 displayed dissimilarity among the groups' science attitude changes toward the science PCA. One would expect a similarity among group profiles if attitude changes were related to membership in the specialization area groups.

Table 13 exhibited similar profiles for language arts, social studies, mathematics, and special education groups. The early childhood and science group profiles did not bear a resemblance to other group profiles for science teaching attitude changes. Four group profiles reflected science teaching attitude changes during both phases of the course and movement toward the PCA. The early childhood group responded more to the on-campus treatment and the science group responded more to the field experience treatment.

These findings demonstrated a relationship between group science teaching attitudes and PCA, but little or no relationship existed between group science attitude changes and PCA. Null Hypothesis 4 was rejected for science teaching attitude changes, but the findings failed to support rejection of the hypothesis for science attitude changes. Content specialization by groups was considered an attitude change factor for science teaching attitudes only.

Perceptions of science training. "High," "above average," "average," and "low" perception groups were formed from STAS attitude scores on the 5-P and 5-N scales. Population means increased on each measure for a growth of 6.12 on a 30 point scale. Perception group profiles were constructed for science (Table 17) and science teaching (Table 18) attitude changes.

"High" and "above average" perception groups reflected attitude changes toward the science PCA level throughout the treatments.

Attitudes of the "average" perception group were unchanged. The "low" perception group consisted of three individuals and experienced science attitude increases through the end of the methods treatment, but not for the field experience.

"High," "above average," and "average" perception group profiles demonstrated science teaching attitude changes toward the PCA with higher perception groups exceeding or more closely approaching the PCA level. The "low" perception group science teaching profile did not consistently approach the PCA.

Relationships between perceptions of science training groups, attitude changes and perceptions of credible communicator attitudes existed among some groups. Groups that maintained "high" or "above average" perceptions exhibited stronger relationships between science and science teaching attitudes and PCA. These findings served to support rejection of Null Hypothesis 4 and made possible the conclusion that higher perception groups, more than lower perception groups, experienced attitude changes that were related to the perceived attitudes of the most credible communicators. Therefore, an individual's perception of science training was considered a factor in the science and science teaching attitude change process.

Science achievement. Individuals were grouped based upon science grade point averages (GPA). Four individuals composed the "high" achievement group, ten were "above average," nine "average," and two individuals comprised the "low" achievement group.

As illustrated in Table 21, achievement group science attitude profiles lacked relationships in movement toward the PCA. However, science teaching attitude change profiles (Table 22) did indicate attitude movement toward the PCA, with "average" and above groups demonstrating closer attainment of the PCA level.

In addition to science achievement, the researcher investigated attitude changes for groups based upon the number of science courses completed before the elementary science methods course. Seven individuals had completed four science courses, nine finished three courses, seven completed two, and two individuals had finished one science course. The latter group contained too few individuals for reliable attitude change information.

Two patterns emerged in the comparison of science attitude change profiles (Table 25). Groups completing one and four science courses demonstrated attitude changes toward the PCA throughout the treatments. Groups with three and two completed science courses revealed science attitude changes toward the PCA after the methods treatment, but these changes were followed by decreased group attitudes after the field experience. In contrast to the above, the patterns of group profiles for science teaching attitude changes were nearly identical and all groups demonstrated attitude growth toward the PCA.

For achievement and completed science course groups, relationships between attitude changes and the level of PCA were more evident for attitudes toward science teaching. Positive relationships existed less frequently for science attitudes. Groups based upon completed science courses maintained stronger relationships. The existence of relationships between group attitude changes and perceptions of credible communicator attitudes were cause for rejection of Null Hypothesis 4, but for science teaching attitudes only. Achievement in science and the number of completed science courses were factors which influenced preservice teacher attitudes toward science teaching more often than changes in attitudes toward science.

Summary. The following items summarize the previous conclusions:

1. Preservice elementary teacher attitudes toward science teaching substantially increased throughout the study, whereas attitudes toward science did not change.

2. Most credible communicator ratings were shared by three principal communicators: science instructor, graduate assistant, and a supervisor/unit evaluator. Overall, the science instructor acquired the highest credibility rating.

3. Preservice elementary teacher attitudes toward science generally increased toward the perceived attitude level of the most credible communicator (PCA), whereas science teaching attitudes clearly changed toward the PCA.

4. All of the respondent factors (content area specialization groups, perceptions of science training, science achievement, and the quantity of completed science courses) were related to the preservice teachers' attitudes toward science teaching. Perception of science training was the only factor found related to attitudes toward science.

5. In general, the results of the study provide evidence that the principle of communicator credibility applied in the context of a team taught preservice elementary teacher science methods course.

Implications

The preceding conclusions are believed to have the following implications for attitude change within the context of teacher-education programs.

Hovland's Credibility Principle

Hovland's communicator credibility principle maintains that a communicator's expertness and trustworthiness influences the respondent's evaluation of the attitude change messages and affects the degree to which messages are accepted and attitudes changed. Basic is the assumption that communicator credibility affects attitude changes of the respondents toward the position advocated by the communicator. In general this study has shown the validity of this assumption through population attitude change profiles in which a major attitude movement occurred at various points throughout the study. As simple as the principle seems, the results were not as simply explained. Because many individuals of the population did experience attitude changes but did not exhibit movement toward the PCA, the researcher suspects that the communicator is not the key agent of attitude change. Furthermore, according to the principle, an individual's attitude should move toward a lower level if the individual perceives the most credible communicator's attitude to be lower. Close scrutiny of individual attitude changes and PCA reveals mixed findings on this item. In the context of this study, the principle was too simple and did not serve as a total explanation of attitude change. In general terms, the principle applied, but better helped to explain attitude changes when coupled with the attitude change findings of additional respondent factors. While credible communicators may not be _the_ key agents of attitude change, the researcher suspects that credibility is _an_ agent of attitude change and is an item useful for improving attitudes in education.

Science Education

Appropriate attitudes toward science and science teaching are important factors in successful science teaching by elementary school teachers. If preservice teacher attitudes toward science and science teaching can be increased to an acceptable level and maintained, then the students of these teachers will be well served through improved teaching behaviors. Hovland's principle suggests that attitude changes result when the respondent is brought into the environment of a credible communicator and the respondents' attitudes will move toward their perceptions of the most credible communicator's attitude level. If this principle holds to a high degree, the training and preparation of highly credible communicators should be encouraged. A respondent's identification of the most credible communicator may depend upon characteristics of the communication environment apart from the communicator, e.g., the respondents' content areas of specialization, grade point averages in science, number of science courses taken, and perceptions of science training. Therefore, the matching of preservice teachers, as classified by their characteristics, with the appropriate communicator is suspected to result in improved science education for the preservice teacher via improved attitudes. Credible communicator characteristics must also be identified and nurtured in an effort to arrange an appropriate match between teachers and communicators.

Credibility Characteristics

That the most credible communicator will have a positive attitudinal effect on preservice teachers through verbal persuasion is an inference made possible by the relationships inherent in profile comparisons of this study. Teacher education programs which desire to improve attitudes should select instructors who are perceived by preservice teachers as being both expert and trustworthy in the content area of desired change. Where team teaching occurs, all communicators should maintain the highest credibility possible to achieve maximum attitude change results. The high credibility ratings shared among several of the teaching team members of this study can be interpreted as evidence that all communicators need expertise and experience in the subject area of desired attitude change.

A study conducted by Shrigley (1976) reported that a credible science methods instructor was perceived as one who:

1. Refers to practical teaching activities in class.
2. Has taught science to children.
3. Assumed responsibility for teaching content.
4. Models teaching modes similar to those proposed for children.
5. Assists science professors in designing science content courses.
6. Counsels student teachers.
7. Assists inservice teachers (p. 451).

However, the following were not found to enhance credibility: (1) teaching general education topics; (2) teaching subjects other than science; (3) conducting research; and (4) authoring content and methods textbooks. Shrigley (1976) suggested that the most credible communicator who might best affect a positive attitude change should be a practitioner, one who could draw upon a store of experiences with the content area, demonstrate several modes of teaching, change theory into practice, and counsel preservice and inservice teachers. What is known about credibility characteristics should guide the selection of a university instructor or members of a teaching team if optimal attitude change results are to be realized.

But, the university must also consider the nature of cooperating school programs and the credibility of the field school's teachers. In effect, a position of attitude change responsibility has been delegated to the classroom teacher once the teacher education program enlists the services and facilities of field schools as practical learning laboratories for its preservice teachers. To provide a successful and supportive environment for the nurturing of attitude changes, it is necessary to ensure that the classroom teacher is able to fulfill her responsibilities. It seems inequitable that cooperating teachers were ranked last in the preservice teacher perceptions of credibility which resulted in this study. Teacher education programs must do more to help the classroom teachers improve their credibility. The results of this study may be viewed as a need for the teacher education program to assess the needs of its students and cooperating teachers and to provide inservice training or program indoctrination as a means of enhancing communicator credibility, maintaining effective communication, and reaping the mutual benefits of increased, positive preservice teacher attitudes. Failure to articulate and coordinate a systematic program of preservice teacher education, which relies upon credible communicators, will result in the education of new teachers who have low affect and insubordinate the school curriculum and schooling.

Teacher Education

In addition to the above, teacher education programs and the communicators used to convey the communications must define the affective behaviors expected of preservice teachers. Daily lessons typically center around a conceptual theme, a major principle, or some other type of cognitive learning outcome. Affective learnings, if considered at all, are considered peripheral to the central theme. However, the processes associated with science and its teaching deserve more attention. Content acquisition occurs in content-centered courses. To be scientific one must not only know about science, one must practice it also. To be scientific means that one has such attitudes as open-mindedness, curiosity, rationality, honesty, suspended judgment, critical mindedness, objectivity, and humility. Therefore, for a preservice teacher to learn the proper attitudes of science and science teaching, that teacher must be provided repeated opportunities to engage those attitudes in the content, methods, and field experience classrooms. The teacher education program truly must be coordinated throughout so that not only are the affective behaviors identified, but also they become reality through practice. To know that ideal affective behaviors exist and are appropriate is not enough. Preservice teachers must know that they are expected to demonstrate those behaviors, and the curriculum, at all levels, must provide adequate and appropriate opportunities for those affective behaviors to be learned.

Adjustments in the curriculum of a coordinated teacher education program or circumvention of its requisites are not likely to result in maximum attitudinal changes. Consider a case in point. Of the preservice teacher factors investigated, perceptions of training, achievement, and number of courses completed in the attitude change subject area were closely related to attitude changes. A relationship appears to exist among the factors in that they all are concerned with program prerequisites. In this study, all preservice teachers were to have completed two of four science courses before enrolling in the science methods course. Sixty-four percent exceeded the requirement. As evidenced by group attitude comparisons, more training in the subject could have increased perceptions of training and may have resulted in improved achievement for the remaining 36% of the population. In a separate study, Gabel (1981) reported that a greater number of science courses completed by students had a positive effect on science attitudes, with four science courses being an optimal prerequisite for a methods class. The findings of this study corroborate Gabel's earlier findings and suggest that teacher education programs which desire to improve attitudes should determine optimal subject area requisites and require compliance before permitting admission to a course where attitude changes are to be emphasized. On the basis of the findings of this study, four content science courses should have been completed before the methods course. The link between the number of courses, perceptions, and attitudes appears to result in increased attitudinal change in a positive direction.

Recommendations

The following recommendations are suggested for replication of the study:

1. Locate and draw upon a population having more individuals who are members of the special factor groups. Some groups in this study had too few members to provide reliable indications of what attitude change profiles could be expected under certain conditions. It is helpful to compare group profiles when the group sizes are larger and equivalent.

2. Test the principle of communicator credibility in other educational settings to establish whether or not the effects are generalizable to all educational environments, or apply only to select settings. What contextual factors affect the efficacy of the principle?

3. Identify additional group factors which may have been overlooked, yet which may also influence attitude change. Consider, for example, social groups or groups formed by preferences of teaching styles. Because the possibility of cumulative and interactive effects exist, study the group factors in a systematic fashion as a part of a longitudinal study. Do the factors apply and do effects on attitudes endure throughout time?

4. Consider individual respondent factors which may have an influence on the processes that affect the internalization of communications. The degree of open and closed-mindedness, variations in age, gender and aptitude may influence the way individuals respond to communications and perceive credibility.

5. Systematically test Hovland's principles to verify that these principles do indeed affect attitude change in teacher education environments. The following questions merit further consideration:

> Are the credibility components of expertness and trustworthiness inseparable in education? Perhaps one of the components more effectively causes attitude changes in an environment where respondents spend frequent and considerable time with a communicator. If so, what programs can help schools enhance the credibility of all teachers?

> Are the effects on attitudes the result of differences in the amount of student attention or comprehension? Are the effects really more closely related to student motivation?

Perhaps the effects result through cognition more frequently than emotional factors such as motivation. What bearing would these findings have on teaching style and instruction?

With students typically having lengthy exposures to teachers of high and low credibility in education, do the effects of credibility disappear after a few weeks as in Hovland's studies? Or, do low credibility effects undo the positive affect brought about by high credibility sources? If so, what implications result for school staffing?

Do differences in student mental abilities influence susceptibility to persuasion and affect attitude change? Are students of lower mental abilities resistant to persuasion, or are they credulous and easier to change? Does the mental ability of a student affect the endurance of any attitude change? What implications exist for curriculum, instruction, and school organization?

6. Although credibility is an important aspect of the attitude change, it is a general aspect of the process. Specific characteristics that make one communicator more credible than another within a particular setting should be identified. Are there similarities among credible communicator characteristics within a variety of settings? What implications result for attitude change due to similarities or dissimilarities?

7. The nature of the communication is suspected to be of equal or greater importance than credibility. Future research should focus on the nature of the message and verbal and nonverbal behaviors of the communicator. Identify the salient aspects of the dynamics of communication exchange and the impact those aspects have on attitude

changes. Identify aspects related to the following factors: content, communicator, media, and situation. What matches result in credibility, communication, and unbound respondent persuasibility factors that involve an individual's general susceptibility to many different types of persuasion and social influences within the context of education?

8. Finally, assuming that attitudes were changed to a desirable level, how can communicators induce resistance to persuasive influences which are counter to the desired changes? What communications or reinforcers are needed to maintain attitude levels throughout the future? How can these reinforcers be built into curriculum and instruction? Perhaps future research will provide answers to these questions.

Concluding Statement

The intent of this study was to identify Hovland's principles of attitude change and test the primary principle of communicator credibility as it might apply to attitude changes in teacher education. The results of this study permit the effects of the principle to be affirmed in a general sense through the existence of relationships between preservice teacher attitude changes, perceptions of credibility, and perceptions of credible communicator attitudes.

In retrospect, if the researcher were to repeat this particular study, attempts would be made to measure individual perceptions of all communicators at posttest and postposttest times. Two measures for each communicator would permit the researcher to determine if changes in attitude perceptions occurred and determine if changes affected the testing of Null Hypothesis 3. Also, if logistics would permit, PCA measurements could occur independently to increase opportunities for valid responses, reduce possible sources of bias, and reduce measurement fatigue.

This investigation took as its foundation principles of attitude change which were believed generalizable to teacher education. The study was limited in scope, however, to the principle of credibility and selected respondent factors. Perhaps future studies will verify that these principles have value to teacher-education and provide implications for schools, teachers, and instruction by testing the principles in other settings and examining additional aspects of the attitude change process as related to credibility, communicators, communications, and learners.

REFERENCES

REFERENCES

Allport, G. W. Attitudes. In C. Murchison (Ed.), A handbook of social psychology. Worcester, Mass.: Clark University Press, 1935.

Ausubel, D. P. How reversible are the cognitive and motivational effects of cultural deprivation? In B. Rosenshine (Ed.), Enthusiastic teaching: A research review, School Review, 1970, 499-515.

Blackwood, P.E. Science in the elementary school. School Life, 1964, 47 (2) 13-15.

Borg, W. R. & Gall, M.D. Educational research. New York: Longman, 1979.

Butzow, J. W. & Davis, A. The development of a semantic differential test of teachers' attitudes toward teaching elementary school science. Science Education, 1975, 59 (2), 211-220.

_____. & Williams, C. The content and construct validation of the academic-vocational involvement scale. Educational and Psychological Measurement, 1973, 33, 495-498.

Bybee, R. W. The ideal elementary science teacher. Paper presented at the National Science Teachers Association Annual Meeting. New York, 1972. (ERIC Document Reproduction Service No. ED 064 117)

Coleman, J. The adolescent society. New York: Free Press, 1961.

Crooks, R. C. The effects of an interracial preschool program upon racial preference, knowledge of racial differences, and racial identification. Journal of Social Issues, 1970, 26 (4), 137-144.

Fishbein, M. & Ajzen, I. Belief, attitude, intention and behavior. Reading, Mass.: Addison-Wesley, 1975.

Gabel, D. Attitudes toward science and science teaching of undergraduates according to major and number of science courses taken and the effect of two courses. School Science and Mathematics, 1981, 81 (1), 70-76.

Gallagher, J. J. & Korth, W. W. Attitudes of seniors concerning science. A paper presented at the Annual Meeting of the National Association for Research in Science Teaching, Pasadena, California, 1969.

Good, T. L., Biddle, B. J., & Brophy, J. E. Teachers make a difference. New York: Holt, Rinehart & Winston, 1975.

Hagerman, B. H. A study of teachers' attitudes toward science and science teaching as related to participation in a CCSS project and to their pupils' perceptions of their science classes. Unpublished doctoral dissertation, Indiana State University, 1974.

Halloran, J. D. Attitude formation and change. Great Britain: Leicester University Press, 1970.

Hamachek, D. Characteristics of good teachers and implications for teacher education. Phi Delta Kappan, 1969, 50, (6), 341-344.

Haney, R. E. The development of scientific attitudes. The Science Teacher, 1964, 31 (8), 33-35.

Hovland, C. I., Janis, I. L. & Kelley, H. H. Communication and persuasion. New Haven: Yale University Press, 1953.

Jenkins, J., Wallace, R. A. & Suci, G. J. An atlas of semantic profiles for 360 words. American Journal of Psychology, 1958, 71, 688-699.

Jingozian, C. R. A study of the relationships between certain teacher practices and student attitudes in the secondary science classroom. Unpublished doctoral dissertation, The University of Texas at Austin, 1973.

Kerlinger, F. N. Foundations of behavioral research. New York: Holt, Rinehart & Winston, 1964.

Kiesler, C. A., Collins, B. E. & Miller, N. Attitude change. New York: John Wiley & Sons, 1969.

Klopfer, L. E. Science education in the 1980's. Science Education, 1980, 64 (1), 1-6.

Loree, M. R. Shaping teachers' attitudes. In B. O. Smith (Ed.), Research in teacher education. Englewood Cliffs, New Jersey: Prentice-Hall, 1971.

Moore, R. W. The development, field test, and validation of scales to assess teachers' attitudes toward teaching elementary school science. Science Education, 1973, 57 (3), 271-278.

_____. The development, field test, and validation of scales to assess teachers attitudes' toward teaching elementary school science. A paper presented to the National Association for Research in Science Teaching, Chicago, 1972.

Morrisey, J. T. An analysis of studies on changing the attitude of elementary student teachers toward science and science teaching. Science Education, 1981, 65 (2), 157-177.

Osgood, C. E., Suci, G. J. & Tannenbaum, P. H. The measurement of meaning. Urbana: The University of Illinois Press, 1971.

Oskamp, S. Attitudes and opinions. Englewood Cliffs, New Jersey: Prentice Hall, 1977.

Petersen, R. & Carlson, G. A summary of research in science education - 1977. Science Education, 1979, 63 (4) 497-501.

Rambally, S. The relationship between teachers' attitudes toward science, verbal interaction patterns, and student achievement in chemical education materials study. Unpublished doctoral dissertation, University of Northern Colorado, 1977.

Riley, J. P. The influence of hands-on science process training on preservice teachers' acquisition of process skills and attitude toward science and science teaching. Journal of Research in Science Teaching, 1979, 16 (5), 373-384.

Rosenshine, B. Enthusiastic teaching: A research review. School Review, 1970, 499-515.

Shaw, M. E. & Wright, J. M. Scales for the measurement of attitudes. New York: McGraw-Hill, 1967.

Sherif, M. & Sherif, C. W. Social psychology. New York: Harper & Row, 1969.

Shrigley, R. L. Credibility of elementary science methods course instructor as perceived by students: A model for attitude modification. Journal of Research in Science Teaching, 1976, 13 (5), 449-453.

_____. The attitudes of pre-service elementary teachers toward science. School Science and Mathematics, 1974, 74, 243-250.

Smith, B. L., Laswell, H. D. & Casey, R. D. Propaganda communication and public opinion. Princeton University Press, 1946.

Stake, R. E., Easely, J. A., et al. Case studies in science education. In L. E. Klopfer (Ed.), Science education in the 1980's. Science Education, 1980, 64 (1), 1-6.

Stern, C. & Keislar, E. R. Teacher attitudes and attitude change: A research review. Journal of Research and Development in Education, 1977, 10 (2), 63-76.

Triandis, H. C. Attitude and attitude change. New York: John Wiley, 1971.

Wagner, R. V. & Sherwood, J. J. The study of attitude change. Belmont, California: Brooks/Cole Publishing Company, 1969.

Washton, N. S. Improving elementary teacher education in science. In D. L. Williams & W. L. Herman Jr. (Eds.) Current research in elementary school science. New York: MacMillan, 1971.

Winer, B. J. _Statistical principles in experimental design_. New York: McGraw-Hill, 1962.

Wish, P. A. _A study of the relationship of science attitudes, selected personal characteristics and situational factors to science teaching behavior of preservice elementary school teachers._ Unpublished doctoral dissertation, North Carolina State University at Raleigh, 1976.

APPENDIXES

Appendix A

The University of Toledo

Competency - Based Teacher Education Program

in

Elementary Education

INTRODUCTION*

The teacher education program at The University of Toledo, is the result of the College of Education having been chosen by the United States Office of Education as one of nine institutions in the nation to design and develop a competency based teacher education program (CBTE) applicable to elementary education.

As a beginning point in the redesign effort in elementary education, a set of general goals for teacher education was adopted to act as guideposts for the development effort. These goals are as follows:

1. Each teacher should be prepared to employ teacher behaviors which will help every child acquire the greatest possible understanding of himself and an appreciation of his worthiness as a member of society.

2. Each teacher should be prepared to employ teacher behaviors which will help every child acquire understanding and appreciation of persons belonging to social, cultural, and ethnic groups different from his own.

3. Each teacher should be prepared to employ teacher behaviors which will help every child acquire to the fullest extent possible for him mastery of the basic skills in the use of words and numbers.

4. Each teacher should be prepared to employ teacher behaviors which will help every child acquire a positive attitude toward school and toward his learning process.

5. Each teacher should be prepared to employ teacher behaviors which will help every child acquire the habits and attitudes associated with responsible citizenship.

6. Each teacher should be prepared to employ teacher behaviors which will help every child acquire good health habits and an understanding of the conditions necessary for the maintenance of physical and emotional well-being.

---

*Excerpt from the Elementary Education 320 module, College of Education, The University of Toledo, Toledo, Ohio. Revised Spring, 1977.

7. Each teacher should be prepared to employ teacher behaviors which will help every child acquire opportunity and encouragment to be creative in one of more fields of endeavor.

8. Each teacher should be prepared to employ teacher behaviors which will help every child understand the opportunities open to him for preparing himself for a productive life and should enable him to take full advantage of these opportunities.

9. Each teacher should be prepared to employ teacher behaviors which will help every child understand and appreciate as much as he can of human achievement in the natural sciences, the social sciences, the humanities, and the arts.

10. Each teacher should be prepared to employ teacher behaviors which help every child to prepare for a world of rapid change and unforeseeable demands in which continuing education throughout his adult life should be a normal expectation.

The second step was the examination of these goals in relation to five primary contexts in order to provide for the transformation of the general goals into more specific objectives. An initial assumption was that five conditions of life and education were of major importance and must be considered in the formulation of a program of teacher education. These conditions, referred to as "contexts," were:

1. Instructional Organization
2. Educational Technology
3. Contemporary Learning-Teacher Process
4. Societal Factors
5. Research

Subsumed under these five contexts are the important sources of change and necessary learning in a program of teacher education. It is for this reason that the student in all elementary education CBTE blocks will find the units of instruction or "modules," representative of these contexts, and in fact, evolved from them. Also, as mentioned previously, these context areas served as foci for the development of more specific statements of behavior relevant to teaching and representative of the general goals of the redesign effort. The student will see the results of this within the program when examination is made of the behavioral or performance objectives. They are statements which further define the goals of the program within each of the five context areas.

As a final note in understanding the background of the current program at Toledo, you should be aware of certain basic assumptions that are implicit in the Toledo model. These are as follows:

> Five conditions of life and education (i.e., contexts) are of major importance to the teacher education.
>
> Modern teacher education must prepare teachers for the schools of the future, which we define as those employing a differentiated staff, team teaching elementary school organization. Students must receive their training under this type of school organization and practice.
>
> All elements or target populations in the educational system must be given appropriate and adequate training and retraining to the best degree possible in each situation, or there will be only limited, negligible change in teacher education curricula and elementary education programs.
>
> New and retrained teachers must receive intelligent and sympathetic support in the schools where they are placed to minimize risks of teacher failure and general educational unresponsiveness to change.
>
> The new program requires the development and utilization of individually guided education.
>
> Elementary teachers shall be prepared to specialize in one field (social studies, reading and language arts, mathematics, early childhood, special education or science) and to generalize in the others. Preschool teachers shall receive general preparation in the subject matter of elementary education.
>
> Teachers shall be instructed by, and taught to use, the most recent technological and media innovations.

The new program shall apply behavioral modification as well as appropriate practices dictated by developmental psychology.

The new program requires an awareness of the differences existing in society today.

Teachers who complete the new program will know how to assess and modify their own teaching behavior and style.

The new program will incorporate various levels of experience (e.g., simulation, games, role playing) in order to approximate more ideally the realities of teaching.

The public schools are a vital part of the teacher education program.

The university must be changed in ways conducive to the needs of the new program, and these changes shall precede all other activities.

Teacher preparation is continuous. The time required for any candidate to complete the training program will depend only on satisfying the performance criteria.

The professional elementary education courses in the CBTE program are as follows:

Course 1 - 312:320 - Elementary Teaching and Learning I

Course 2 - 312:324 - Elementary Teaching and Learning II

Course 3 - 312:328 - Elementary Teaching and Learning III

Course 4 - 312:340 - Elementary Teaching and Learning IV

Course 5 - 312:392 - Student Teaching

Courses 320 through 340 are sequential in nature and deal with training the preservice elementary education teacher with specific competencies within each of the 5 context areas and that relate to the general goals of the CBTE program at Toledo. Student Teaching, 392, is a synthesis of these experiences that is completely field based in the participating elementary schools in which the student demonstrates and practices the skills achieved in the prior courses.

Course 1 - 312:320 - Elementary Teaching and Learning I - Mathematics

There are four major directions to the instructional objectives included within the 320 course. These directions are as follows:

1. Introduction to interpersonal communications and its value within instruction.

2. Introduction to the organization and operation of elementary schools.

3. Introduction to the processes and procedures of planning for instruction in the elementary school.

4. Introduction to mathematics education processes and procedures in the elementary school.

There are nine instructional modules that students complete to experience the four major directions of the 320 course. They are:

1. The Elementary School Mathematics Program
2. Individually Guided Education and its Seven Components
3. Writing Performance Objectives
4. Systematic Planning for Instruction
5. Components of Elementary Mathematics Instruction
6. Materials for Mathematics Instruction
7. Sequence of Instruction -- Elementary School Mathematics
8. Group Process in Education
9. Implementing Instruction - Mathematics

In addition to campus-based instruction, teacher education students receive instruction and apply their knowledge in real-life elementary school settings. During the field application portion of 320, students spend 55 hours in schools throughout the academic quarter.

Course 2 - 312:324 - Elementary Teaching and Learning II - Language Arts

312:324 is designed to include elements of educational psychology and language arts, and a credit/no credit module for media technology.

Given a forty (teaching) day quarter, the preservice teachers follow a sequence that calls for 8 classroom days, 3 field days, 14 classroom days, 13 field days, and 2 classroom days. Students are expected to acquire concepts necessary for effective functioning as teachers and are expected to demonstrate these concepts both at the literal (planning) level and at the application (teaching) level before competence may be presumed by the instructional team.

Ten modules comprise the instruction. In addition, students spend 55 hours in field schools applying the on-campus instruction. The modules include:

1. Handwriting Skills
2. Teaching Reading: From Theory to Practice
3. Organizing Instruction to Meet Individual Needs
4. Motivating and Improving Language Growth
5. Effecting Teaching Strategies in Language Arts
6. Learning to Make the Distinction Between the Identification of a Behavior and the Formation of an Inference
7. Recording Observed Behavior and Forming Summaries Based on the Behavior
8. Strategies for Changing Behavior
9. Cognitive Developmental Periods in Children
10. Understanding and Operating Basic Audiovisual Equipment

Course 3 - 312:328 - Elementary Teaching and Learning III - Science

The 328 block has four areas of emphasis: the systematic planning of whole units of instruction, the design and production of high quality instructional materials, the acquisition of a variety of teaching techniques, and a strong focus on teaching science.

There are nine modules in the 329 block:

*1. Unit Planning and Implementation

*2. Teaching Science in the Elementary School

*3. Selection and Use of Instructional Media

*4. Basics of Preparing Visual Instructional Materials

5. Problem Solving

*6. Classroom Management Techniques

*7. Concept Lessons

*8. Inquiry Teaching

*9. Questioning Lessons

Eight of these modules include a field experience component and involve the use of a checklist to gauge the students' performance. Preservice teachers are scheduled for 3 1/2 to 4 weeks of field experience (plus a number of days early in the quarter) to do the following:

1. teach science.

2. observe and analyze other participants' teaching.

3. plan and prepare for teaching.

4. assist the cooperating teacher.

---

*Include Field Experience Checklists, see Appendices C and D.

Course 4 - 312:340 - Elementary Teaching and Learning IV - Social Studies

The 340 block is divided into two parts: (1) a five week segment (on-campus) concerned with theory and knowledge instruction and, (2) application of campus instruction while teaching self-constructed modules on social studies topics for five weeks in an elementary classroom. Instructional modules include:

1. Instructional Design of Educational Simulation Games
2. Social Studies Planning
3. Values Clarification
4. Role Behavior of Teachers on Teams
5. Teacher Professionalism
6. Module Planning and Evaluation
7. Developing the Interdisciplinary Unit
8. Module Implementation

Appendix B

Elementary Teaching and Learning III

312:328

Course Calendar

Winter Quarter, 1981

Elementary Education 312:328
Course Calendar
Winter Quarter, 1981

Module Instruction Phase

Week	Monday	Tuesday	Wednesday	Thursday
1	1/5 Open Registration	1/6 SM 216 Orientation, School Groups	1/7 FIELD Scavenger Hunt Day 1	1/8 SM 216 Unit Planning Overview
2	1/12 SM 377 Science	1/13 SM 216 Inquiry Lessons	1/14 SM 216 Questioning; Classroom Management	1/15 SM 216 Unit Planning; Unit Lab ---------- SM 377 Science
3	1/19 Martin Luther King Day (Holiday)	1/20 TEC Selection and Use of Instructional Media	1/21 TEC Special Education	1/22 TEC Special Education ---------- Unit Planning; Unit Lab
4	1/26 SM 377 Science	1/27 SM 216 Concept Lesson	1/28 FIELD Scavenger Hunt Day 2	1/29 TEC Special Education ---------- Unit Lab
5	2/2 SM 377 Science	2/3 TEC Materials Production Day 1	2/4 TEC Materials Production Day 2	2/5 SM 216 Unit Lab ---------- Science SM 377

Week	Monday	Tuesday	Wednesday	Thursday
6	2/9 TEC Science	2/10 SM 216 Problem Solving	2/11 TEC Science ---------- Unit Lab TEC	2/12 FIELD
7	2/16 TEC Unit Lab	2/17 SM 377 Science	2/18 F I E L D E X P E R I E N C E	2/19
8	2/23 F I E L	2/24 D E X P	2/25 E R I E	2/26 N C E
9	3/2 F I E L	3/3 D E X P	3/4 E R I E	3/5 N C E
10	3/9 F I E L D	3/10 E X P E R I	3/11 E N C E	3/12 Science Fair (2 ideas from each student; Course Evaluation
11	3/16 Other post- assessments; Evaluation Feedback	3/17 Faculty in their offices	3/18 Final grades recorded.	3/19

APPENDIX C

The University of Toledo

Elementary Teaching and Learning III

312:328

Module 02

Teaching Science In The Elementary School

Winter, 1981 Revision/Martin

Elementary Education

312:328

Module 2

I. **Department/Context**: Instructional Organization

II. **Topic**: Necessary Science Teaching Methods for Instruction

III. **Title**: Teaching Science in the Elementary School

IV. **Prerequisites**: Completion of 312:320, 324 and a minimum of eight hours of undergraduate science.

V. **Rationale**:

Improvements in the teaching of science will make a difference in the total education and lives of students and lead to the improved quality of society. Hence, "the major goal of science education is to develop scientifically literate and personally concerned individuals with a high competence for rational thought and action" (Carin & Sund, 1980, p. 39). This is possible through the development of values, attitudes, and inquiry skills.

As a future teacher, your attitude toward science and its teaching is critical to the attitudes and achievements of your students. Your experiences with science most likely will influence the way you teach it to your students. The methodology of this module is based upon Piagetian principles which, in essence, are represented by the following Chinese proverb:

> I Hear and I Forget
> I See and I Remember
> I Do and I Understand

VI. **Objectives**:

TPO 1. The preservice elementary teacher shall participate in "hands-on" science inquiry activities and apply the "hands-on" approach in the field setting by:

EO 1-1. Participating in all selected activities from ESS, SCIS, and SAPA during the science methods class and completing the following readings: DeBruin, Ch. 10; Carin & Sund, Ch. 8.

EO 1-2. Selecting and applying appropriate instructional science methodologies as typified by elementary science programs when constructing and teaching the science unit. At least 50% of the unit lessons shall consist of "hands-on" activities.

TPO 2.   The preservice elementary teacher shall participate in and demonstrate his/her teaching competence in science processes through:

    EO 2-1.   Unit plan activities which shall contain, properly labeled, the following science processes* (Reading: Carin & Sund, Ch. 1, 3):

        2-1 a. Observing
            b. Classifying
            c. Measuring
            d. Hypothesizing or Predicting
            e. Describing
            f. Inferring or making Conclusions from data
            g. Asking insightful Questions about nature
            h. Formulating Problems
            i. Designing Investigations including Experiments
            j. Carrying out Experiments
            k. Constructing from data, Principles, Laws, and Theories

    *Each process skill shall be used at least once during the teaching of the unit.

    EO 2-2.   Cognizance and demonstration of the following scientific attitudes (Reading: Carin & Sund, Ch. 1):

        2-2 a. Curiosity
            b. Humility
            c. Skepticism
            d. Open-mindedness
            e. Avoidance of dogmatism or gullibility
            f. Positive approach to failure
            g. Objectivity

TPO 3.   The preservice elementary teacher shall: (1) develop an appreciation for the historical, philosophical, and current significance of science to society; and (2) develop positive attitudes about science and science teaching.

    EO 3-1.   Appreciation for the historical, philosophical, and current significance of science to society shall be developed through class instruction and completion of the following readings:

        3-1 a. DeBruin, Chapter 10
            b. Carin & Sund, Ch. 1, 3, 4, 8, Appendix A
            c. Abruscato and Hassard (Loving and Beyond) Ch. 1, 4, 5

    EO 3-2.   The development of positive attitudes toward science and science teaching shall be measured by the following instruments:

3-2 a. Science Teaching Attitude Scales, Pretest, Posttest, and Postposttest
b. Perceptions of Communicator Credibility, Posttest and Postposttest
c. Perceptions of Communicator Attitudes, Posttest and Postposttest
d. Science Teaching Log (Supplement 2)

TPO 4. The preservice elementary teacher shall participate in science methods activities designed to assist him/her in the production of the science unit, peer teaching lesson, and science fair projects by completing the following:

EO 4-1. Design an instructional science unit according to the criteria of Module 1, Unit Planning and Implementation

EO 4-2. In addition to the above, the following are required for satisfactory completion of EO 4-1.

4-2 a. Construct at least one Science Learning Center which incorporates the guidelines of Module 1, Supplement 1 and suggestions from science methods instruction (Reading: DeBruin, p. 10-15).
b. Identify and apply safety procedures in daily lesson planning and instruction (Consult Module 1, Supplement 3. Read: DeBruin, Ch. 3).

EO 4-3. Select and condense into a mini lesson* (approximately 5 minutes), one daily lesson plan, and present to peers during science methods class. The lesson shall demonstrate one of the following:

4-3 a. Science Concepts, as per checklist A-31.**
b. Convergent Inquiry, as per checklists A-33-36.**
c. Divergent Inquiry, as per checklists A-37-40.**
d. Questioning, as per checklist A-41.**

*The lesson is to be team taught (Read: Carin & Sund, Ch 6).

**Checklists are located in Supplement 11, pp. 2-54-61.

EO 4-4. In addition to cognitive unit evaluation, construct and administer the following evaluation components: (Reading: DeBruin, Ch. 2)

4-4 a. Affective evaluation of three self-selected daily lessons.
b. Affective evaluation of the learning center(s).
c. Affective evaluation of the unit.
d. Devise a record-keeping system containing cognitive, affective and psychomotor information for each student.
e. Daily Self-Evaluation (Supplement 3)

EO 4-5. Select* two projects from your unit for display and discussion during the end of the quarter Science Fair. Projects must be copied onto a ditto, available in classroom quantity, and may consist of the following:

4-5 a. Learning Center ideas
b. Bulletin Board ideas
c. Science activities: manipulative, critical-thinking, reading improvement, values clarification, community projects, etc.
d. Record-Keeping systems
e. Science games, flash cards, etc.
f. Unique Evaluation instruments
g. Inexpensive, self-constructed science equipment
h. Other.

*Each preservice elementary teacher must select, copy, and display two projects.

TPO 5. The preservice elementary teacher shall demonstrate skills for effective human relations with classroom children, team member, and cooperating teacher through:

EO 5-1. Recognition of the importance of individual children and demonstration of confidence and flexibility in relations with children, team member, and cooperating teacher. To be measured by:

5-1 a. Team Member Checklist (A-9)
b. Personal and Professional Fitness Checklist (A-11-15)

VII. Pretest: Science Teaching Attitude Scales, Moore, 1973, a 70 item attitude toward science and science teaching inventory.

VIII. Learning Activities: Assigned readings, class activities, unit planning, science fair projects, and selected science activities from ESS, SCIS, and SAPA as listed in Objectives. Examples are contained in the supplement of this module.

IX. Posttest: EO 3-2 and 5-1. Additional assessment through examination of completed activities and teaching of unit.

X. Required Test:

DeBruin, Jerry. Creative, Hands-On Science Experiences. Carthage, IL: Good Apple, 1980.

XI. Recommended Texts:

Abruscato, Joe and Jack Hassard. Loving and Beyond, Science Teaching for the Humanistic Classroom. Santa Monica, California: Goodyear, 1976.

Abruscato, Joe and Jack Hassard. The Whole Cosmos Catalog of
Science Activities. Santa Monica, California: Goodyear, 1977.

Carin, Arthur A. and Robert B. Sund. Teaching Science Through
Discovery, Fourth Edition. Columbus, OH: Charles E. Merrill
Publishing Company, 1980.

XII. Readings:

REQUIRED

DeBruin, Jerry. Creative, Hands-On Science Experiences. Carthage,
IL: Good Apple, 1980. Chapters 2, 3, 10.

Carin, Arthur A. and Robert B. Sund. Teaching Science Through
Discovery, Fourth Edition. Columbus, OH: Charles E. Merrill
Publishing Company, 1980. Chapters 1, 3, 4, 6, 8 and Appendix A.

RECOMMENDED

Abruscato, Joe and Jack Hassard. Loving and Beyond, Science Teaching
for the Humanistic Classroom. Santa Monica, California: Goodyear,
1976. Chapters 1, 4, 5.

Barnes, Carol P. Questioning Strategies to Develop Critical
Thinking Skills. ERIC Document No. 169486.

Carin and Sund, 1980. Chapters 2, 5, 7, 9, 12.

DeBruin, 1980. Chapters 1, 4-9.

DeVito, Alfred and Gerald H. Krockover. Creative Sciencing: A
Practical Approach, Second Edition. Boston: Little, Brown
and Company, 1980. Chapters 1-6.

Gega, Peter C. Science In Elementary Education, Third Edition.
New York: John Wiley & Sons, 1977. Chapters 3, 4, 5.

Kondo, Allan K. "Children Can't Think," Science and Children,
October, 1969, pp. 31-33.

Martin, Ralph E. How To Help 'em Read Better and Still Teach
Science. Unpublished document presented at Toledo IRA
Conference, February, 1978.

_____. Using Literature to Teach Science, 1980 (ED No.:
193 072).

_____. Critical Thinking in the Middle School Science Classroom.
Unpublished paper from National Middle School Association
Annual Conference, November, 1980.

Orrach, Lawrence P. Teaching Science in the Elementary Classroom.
Unpublished Master of Education Project, The University of
Toledo, 1980.

XIII. Bibliography

For a bibliography of useful science references additional to those listed above, see DeBruin, 1980, pp. 217-221.

XIV. Supplements:

1. Science Methods Module Checklist 2-7
2. Science Teaching Log 2-9
3. Daily Self-Evaluation Format 2-10
4. The Elementary Science Study (ESS) 2-11
5. ESS/Special Education Distribution Chart 2-12
6. Science - A Process Approach (SAPA) 2-13
7. Science Curriculum Improvement Study (SCIS) 2-19
*8. ESS Activities . . . . 2-20 - 37
*9. SAPA Activities . . . . 2-38 - 48
*10. SCIS Activities . . . . 2-49 - 53
*11. Checklists for Peer Teaching, EO 4-3 . . . . 2-54 - 61

*Copies available upon request.

Supplement 1

Science Methods Module Checklist

For the use of the preservice elementary teacher. Have you completed the following:

Attendance/Participation

	YES	NO
Attended all science methods classes		
Participated in discussions and activities		

Readings

	YES	NO
DeBruin, 1980 Chapter 2 (EO 4-2, 4-4)		
3 (EO 4-2)		
10 (EO 1-1, 3-1)		
Carin & Sund, 1980 Chapter 1 (EO 2-1, 2-2, 301)		
3 (EO 2-1, 3-1)		
4 (EO 3-1, TPO 5)		
6 (TPO 4)		
8 (EO 1-1, 3-1)		
Appendix A (EO 3-1)		

Unit

	YES	NO
50% of lessons are "hands-on" (EO 1-2)		
Selection and application of current science teaching methodologies (EO 1-2)		
Use of science processes (EO 2-1)		
Science Learning Center (EO 4-2)		
Safety Instructions (EO 4-2)		
Unit Evaluation (EO 4-4)		
4-4 a. Affective evaluation of three self-selected daily lessons		
b. Affective evaluation of learning center		
c. Affective evaluation of unit		
d. Record-keeping system		

	YES	NO

    e. Daily self evaluation . . . .

Science Fair
Display and discussion of <u>two</u> projects (EO 4-5) . . . . . .

Teaching
Demonstration of scientific attitudes (EO 2-2) . . . .

Peer teaching mini-lesson (EO 4-3) . . . .

Human Relations (EO 5-1)

   5-1 a. Team Member Checklist (A-9) . . . .

       b. Personal and Professional Fitness Checklist (A-11-15) . . . .

Assessment (EO 3-2)
Science Teaching Attitude Scales, Pretest . . . .

                                                Posttest . . . .

                                                Postposttest . . . .

Perceptions of Communicator Credibility, Posttest . . . . .

                                                Postposttest . . .

Perceptions of Communicator Attitudes, Posttest . . . . . .

                                                Postposttest . . . .

Science Teaching Log . . . .

Supplement 2

Science Teaching Log

Rationale:

Your attitudes affect the way people respond to you, in general, and specifically, influence the attitudes and achievements of your students. Therefore, it is important to be able to identify occurrences which may affect your attitude toward teaching -- in this block, science teaching.

Purpose:

The purpose of the science log is to provide you a vehicle through which you can recognize and express your feelings and observations as you teach science. You are to maintain a daily written record which is to be submitted to the instructor at the end of the quarter. All logs shall be treated confidentially and shall be read by only the instructor after the science methods module grades have been submitted.

Contents:

Although you will judge what your log's contents shall be, it should contain written expressions of your daily interactions with students, team member, cooperating teacher and university supervisor. Identify the positive and negative occurrences of your daily field experience. What left you feeling most happy and most disturbed? Who said or did what to whom and what were the effects? Consider and try to respond to all of the following questions when writing your log. Also, try to write your log at the end of the day. Your memory will be most complete shortly after the experience has occurred.

How much time did you spend teaching today?
What was the nature of your lesson? Was it successful? Why?
How much time did your team member teach today?
Did you team or turn teach? Were you successful? Why?
Are you confident, interested, optimistic and personally involved with
    science teaching? Why?
Are you a good teacher? Why?
What do you need to become a better teacher?
How much time did your cooperating teacher and university supervisor
    spend with you today? What did they do?
Did they offer feedback? What was the nature of the feedback?
Were they helpful? How?
What kinds of questions did they ask? When did they ask them? What
    was the nature of their questions?
What kind of suggestions did they offer about practical teaching activities?
    About science content?
Have they taught science to children? If so, what kind of teaching model
    are they?
When do they offer counsel?
Are they interested in helping you? Why?
Are they interested in science? Why?
Are they confident, optimistic, and personally involved with your field
    experience?
Are they properly prepared to help you? Why?
What do they need to do to better help you?

Supplement 3

Daily Self-Evaluation Format

Name/Date: _____

CRITERIA FOR SELF EVALUATION*

A. MY GENERAL RATING OF THE LESSON

Circle One: 5 4 3 2 1

(5-Excellent, 3-Acceptable, 1-Unacceptable)

B. I BASED MY RATING ON THE FOLLOWING EVIDENCE

1. Specific STRENGTHS of the lesson

2. Specific WEAKNESSES of the lesson

3. Specific SUGGESTIONS FOR IMPROVEMENT of the lesson (If I could teach it again, how would I change it?)

4. Specific MODIFICATIONS of tomorrow's plan (On the basis of what I learned today and what the students learned, how should I adapt my plan for tomorrow?)

5. Other

*TO BE COMPLETED SHORTLY AFTER LESSON HAS BEEN TAUGHT. SUBMIT TO SUPERVISOR

Appendix D

Elementary Teaching and Learning III

312:328

Course Evaluation Questionnaire

312:328
Course Evaluation Questionnaire

This is a survey asking for your candid opinions about certain aspects of the 312:328 block. The data we receive will help us arrange 328 better in the future. Your identity is to be unknown, but we do ask that you indicate who your facilitator has been this quarter so that data by school groups (but not individuals) can be forwarded appropriately.

For each question below, you are to mark your rating - 1, 2, 3, 4, or 5 - on the answer sheet. Your rating should indicate your feelings - positive to negative - about a certain feature of the program. The scale is as follows:
1 = ++ = strongly, positive, very good, much more, etc.
2 = + = positive, good, more, etc.
3 = 0 = no opinion or mixed feelings
4 = - = negative poor, less, etc.
5 = -- = strongly negative, very poor, much less, etc.

Feel free to write additional comments on these pages for any item.

General

1. Rate how well the 320 Block prepared you for 328.

2. Rate how well the 324 Block prepared you for 328.

3. How would you rate the amount of responsibility placed on you in 328, as compared to earlier blocks?

4. How would you quantitatively rate the work you've done in the 328 Block, as compared to earlier blocks?

5. How would you qualitatively rate the work you've done in 328 Block, as compared to earlier blocks?

6. To what extent did the 328 module instructors demand both quantity and quality of effort from you?

7. How would you rate the quality of feedback (e.g., to the point, helpful, etc.), you've received from your 328 module instructors, as compared to earlier blocks?

8. How would you rate the quantity of feedback (e.g., sufficient, insufficient) you've received from your 328 module instructors, as compared to earlier blocks?

9. How would you rate the difficulty of the 328 block, as compared to earlier blocks?

10. How did you rate yourself as a prospective teacher when the quarter began?

11. How do you rate yourself as a prospective teacher?

　　On the attached sheet, please comment on whether there were specific knowledges or skills that you think you should have encountered/mastered in earlier blocks.

Content

　　How would you rate the <u>value</u> of the topics covered in the modules?

Module #
12. (2) science springboard activities
13. (2) science teaching activities
14. (2) science notebook
15. (2) science Fair
16. (1) writing behavioral objectives
17. (3) systematic selection and evaluation of media for instruction

18. (4) visual materials production processes

19. (8) convergent inquiry skills
20. (8) divergent inquiry skills
21. (8) role of the teacher in inquiry teaching

22. (9) planning and teaching a questioning lesson
23. (7) concept learning and concept teaching

24. (5) skills of problem solving

25. (6) using positive reinforcement and avoiding criticism

26. (1) unit planning
27. (1) unit implementation
28. (10) Special Education Module

Please rate the ways the topics were handled according to the following criteria:

Motivation: Was the topic presented in an interesting way?
Data for Instruction: Was enough information available for you to refer to - such as checklists, models, examples?
Instruction: Was the topic presented clearly and directly?
Feedback of Results: Was constructive feedback given to help overcome problems or to confirm adequate performance?

Note that the number in the box (e.g., #29, for the Motivation element regarding the presentation on "science teaching activities") refers to the number on the answer sheet where you should mark your response.

	MOTIVATION	DATA	INSTRUCTION	FEEDBACK
science teaching activities	29	30	31	32
Science fair activities	33	34	35	36
systematic selection and evaluation of media	37	38	39	40
writing behavioral objectives	41	42	43	44
visual materials production processes	45	46	47	48
convergent inquiry skills	49	50	51	52
divertent inquiry skills	53	54	55	56
role of the teacher in inquiry teaching	57	58	59	60
concept learning and concept teaching	61	62	63	64
skills of problem solving	65	66	67	68
using positive reinforcement and avoiding criticism	69	70	71	72
unit planning	73	74	75	76
unit implementation	77	78	79	80
special education module	81	82	83	84

How would you rate the clarity of the assignments for each module?

85. (2) Science teaching
    (class activities, lab activities, science notebook)

86. (3) Selection and Utilization of Instructional Media
    (media selection process description, using media in instruction)

87. (4) Basics of preparing visual instructional materials
    (designing and producing materials for instruction)

88. (8) Performance Skills in Inquiry
    (convergent and divergent inquiry methods, peer evaluation, self evaluation)

89. (9) Asking Higher Level Questions
    (questioning lesson plan, peer evaluation)

90. (7) Teaching Concept Lessons

91. (5) Problem Solving
    (solving a set of problems)

92. (6) Classroom Management
    (use of positive reinforcement and avoiding criticism)

93. (1) Unit Planning and Implementation

94. (10) Special Education Module

*On the attached sheet, please comment on how your module instructors could be more effective in helping students master the objectives of each of the modules.

312:328
FIELD EXPERIENCE EVALUATION

Below are a series of questions. Please select the answer you think best describes your perception and make your rating on the op scan sheet. (Continue at #95 on the sheet.) Your responses will be anonymous.

1. (95) To what extent did you need help in working with your cooperating teacher(s) and adjusting to your field setting?

2. (96) Rate your facilitator's effectiveness in helping you to adjust to your field setting and cooperating teacher(s).

   * On the attached sheet, please comment on how facilitators could be more effective in helping participants adjust to your field setting and to your cooperating teacher(s).

3. (97) How well informed was your facilitator regarding the instructional modules you had to complete in the field?

4. (98) Rate your facilitator's effectiveness in helping you to meet module requirements.

   * On the attached sheet, please comment on how facilitators could be more effective in helping participants to meet module requirements.

5. (99) To what extent did other participants help you meet module requirements?

6. (100) How well informed was/were your cooperating teacher(s) regarding the instructional modules?

7. (101) To what extent did your cooperating teacher(s) help you meet the module requirements (other than the field checklists)?

8. (102) To what extent did your cooperating teacher(s) help you meet the field experience requirements (especially checklists)?

9-15. How important do you feel it is to have the following people observe you in your field experience? Rate each person below.

   9. (103) cooperating teacher(s)
   10. (104) instructor of module on campus
   11. (105) facilitator
   12. (106) team leader
   13. (107) principal of school
   14. (118) peer team member (participant)
   15. (119) other participant(s)

16. (110) To what extent did you really need observation of your teaching?

162

17. (111) How much time did your facilitator spend observing you per week (with or without using checklists)?

   1. more than 60 minutes per week
   2. 45-60 minutes per week
   3. 30-45 minutes per week
   4. 15-30 minutes per week
   5. less than 15 minutes per week

18. (112) How adequate was that amount (above) of observation time for you?

19. (113) How often was your facilitator at your school?

   1. every day of field experience
   2. averaging three times per week
   3. averaging twice per week
   4. averaging once per week
   5. averaging less than once per week

20. (114) How adequate was the frequency of your facilitator's visits to your school?

21. (115) To what extent did your observing other participants' teaching help you to teach better?

22. (116) To what extent did you perceive a need for feedback on your teaching from your facilitator?

23. (117) Rate the amount of time your facilitator spent in feedback sessions (with you individually and/or with all the participants together):

   1. more than 60 minutes per week
   2. 45-60 minutes per week
   3. 30-45 minutes per week
   4. 15-30 minutes per week
   5. less than 15 minutes per week

24. (118) How adequate was the amount of seminar and/or feedback time?

25. (119) How do you feel about teaming with other students in the field?

   * On the attached sheet, please comment on specific strength and/or weaknesses of your teaming experience.

26. (120) How did you feel about working with your cooperating teacher(s)?

27. (121) How did you feel about your field placement school?

   1. Outstanding! I definitely would like to student teach there.

2. Good. I would accept a student teaching assignment there.
3. Neutral. Makes no difference.
4. I would prefer a student teaching assignment elsewhere.
5. I definitely would not like to student teach there.

* On the attached sheet, please give us any suggestions or comments you have to help us improve the CBTE program. How can it be made better?

312:328
COURSE EVALUATION QUESTIONNAIRE RESULTS

ITEM	RESPONSE (Percentages)				
General	++	+	0	-	--
1.	12	42	23	19	4
2.	8	23	23	38	8
3.	69	27	0	4	0
4.	69	23	8	0	0
5.	92	4	4	0	0
6.	65	35	0	0	0
7.	77	23	0	0	0
8.	69	31	0	0	0
9.	50	46	4	0	0
10.	8	58	27	8	0
11.	58	42	0	0	0
Content					
12.	35	58	8	0	0
13.	58	42	0	0	0
14.	20	40	40	0	0
15.	58	35	8	0	0
16.	8	46	35	12	0
17.	4	31	27	35	4
18.	23	46	19	8	4
19.	23	62	8	8	0
20.	23	69	8	0	0
21.	31	65	4	0	0
22.	31	58	12	0	0
23.	27	54	19	0	0
24.	15	62	8	15	0
25.	58	35	8	0	0
26.	39	54	8	0	0
27.	58	35	8	0	0
28.	8	27	23	31	12

ITEM	RESPONSE (Percentages)				
Topics	++	+	0	-	--
29.	62	38	0	0	0
30.	50	46	4	0	0
31.	62	31	8	0	0
32.	65	27	8	0	0
33.	50	46	4	0	0
34.	54	38	8	0	0
35.	42	42	15	0	0
36.	54	31	15	0	0
37.	4	27	31	35	4
38.	12	38	38	12	0
39.	8	35	31	23	4
40.	12	38	31	15	4
41.	4	38	35	19	4
42.	19	46	35	0	0
43.	12	42	42	4	0
44.	27	35	35	4	0
45.	19	46	8	27	0
46.	31	35	27	8	0
47.	35	38	15	12	0
48.	19	46	15	15	4
49.	42	54	4	0	0
50.	35	62	4	0	0
51.	31	58	4	8	0
52.	42	46	8	4	0
53.	42	54	4	0	0
54.	38	54	8	0	0
55.	38	50	8	4	0
56.	35	62	0	4	0
57.	42	31	27	0	0
58.	19	65	15	0	0
59.	38	50	12	0	0
60.	31	54	15	0	0
61.	23	58	19	0	0

ITEM	RESPONSE (Percentages)				
Topics, continued	++	+	0	-	--
62.	27	54	19	0	0
63.	35	50	15	0	0
64.	27	58	15	0	0
65.	46	19	31	4	0
66.	31	42	23	4	0
67.	31	46	15	8	0
68.	31	35	23	12	0
69.	50	31	19	0	0
70.	38	54	8	0	0
71.	46	50	4	0	0
72.	35	50	15	0	0
73.	27	46	27	0	0
74.	42	38	15	4	0
75.	42	38	12	8	0
76.	54	38	8	0	0
77.	65	23	12	0	0
78.	69	27	4	0	0
79.	69	23	8	0	0
80.	81	15	4	0	0
81.	4	31	35	15	15
82.	8	35	31	19	8
83.	8	42	35	12	4
84.	4	4	19	27	40

Content

ITEM	++	+	0	-	--
85.	58	38	4	0	0
86.	15	27	35	19	4
87.	50	31	15	4	0
88.	27	65	8	0	0
89.	19	69	12	0	0
90.	27	54	19	0	0
91.	15	50	23	12	0
92.	38	46	15	0	0
93.	35	54	12	0	0
94.	8	31	23	19	19

ITEM	RESPONSE (Percentages)				
Field Experience	++	+	0	-	--
95.	15	23	19	19	23
96.	69	15	11	4	0
97.	62	23	8	8	0
98.	69	15	15	0	0
99.	12	50	23	12	4
100.	15	38	19	23	4
101.	15	38	12	35	0
102.	46	31	15	8	0
103.	65	23	12	0	0
104.	23	42	27	8	0
105.	96	4	0	0	0
106.	38	19	38	4	0
107.	12	27	42	15	4
108.	69	19	4	4	4
109.	38	27	35	0	0
110.	35	46	15	4	0
111.	50	27	12	4	0
112.	81	15	4	0	0
113.	88	12	0	0	0
114.	88	12	0	0	0
115.	15	46	35	4	0
116.	50	35	8	4	4
117.	50	8	23	15	4
118.	65	19	12	4	0
119.	54	31	12	0	4
120.	65	19	12	4	0
121.	58	27	12	4	0

APPENDIX E

<u>Science Teaching Attitude Scales</u> (STAS)

also known as

What Is Your Attitude Toward Science And Science Teaching?

by

Richard Moore, 1973

WHAT IS YOUR ATTITUDE TOWARD SCIENCE AND SCIENCE TEACHING?

Richard W. Moore

There are some statements about science and science teaching on the next pages. Some statements are about a person's feelings about science. Some of these statements describe views about how science should be taught. You may agree with some of the statements and you may disagree with others. That is exactly what you are asked to do. By doing this, you will show your attitudes toward science and science teaching.

After you have carefully read a statement, decide whether you agree or disagree with it. If you agree, decide whether you agree mildly or strongly. If you disagree, decide whether you disagree mildly or strongly. Then, find the letter of that statement on the answer sheet, and blacken the space.

- a   if you agree strongly
- b   if you agree mildly
- c   if you disagree mildly.
- d   if you disagree strongly

Example:

00. I would like to have many friends

00.   a▬▬   b▬▬   c▬▬   d▬▬

(The person who marked this example agrees strongly with the statement, "I would like to have many friends.")

Please respond to each statement and blacken only <u>one</u> space for each statement.

Please do not make any marks on this test booklet.

WHAT IS YOUR ATTITUDE TOWARD SCIENCE AND SCIENCE TEACHING?

1. It is important for children to learn that the air is approximately 20% oxygen--at least by the sixth grade.

2. There is no need for the public to understand science in order for scientific progress to occur.

3. Most children should be able to design experiments--at least by the sixth grade.

4. Most people are not able to understand the work of science.

5. When something is explained well, there is no reason to look for another explanation.

6. A teacher should be a resource person rather than an information-giver in science.

7. The products of scientific work are mainly useful to scientists; they are not very useful to the average person.

8. I do not understand science, and I do not want to teach it.

9. A scientist must be imaginative in developing ideas which explain natural events.

10. After all is said and done, it is really the teacher who tells the children what they have to learn and know.

11. Some questions cannot be answered by science.

12. In teaching science, a teacher might spend more time listening to the children than talking to them.

13. Before one can do anything in science, he must study the writings of the great scientists.

14. Rapid progress in science requires public support.

15. Process skills are very important things to be developed in science.

16. Scientists believe that nothing is known to be true with absolute certainty.

17. A major purpose of science is to help man live more comfortably.

18. A new theory may be accepted when it can be shown to explain things as well as another theory.

19. Children must learn certain basic facts in elementary science so they can do well in science in junior high.

20. Scientists do not need public support, they can get along quite well without it.

21. I understand science and I want to teach it.

22. Every citizen should understand science because we are living in an age of science.

23. Children must be told what they are to learn if they are to make progress in science.

24. Science is so difficult that only highly trained scientists can understand it.

25. A teacher has a responsibility to teach the basic processes of science.

26. His senses are one of the most important tools a scientist has.

27. Science may be described as being primarily an idea-generating activity.

28. Ideas are one of the more important products of science.

29. As children experiment, a teacher may give helpful hints, but not the answer to a problem.

30. Science is pretty easy to understand.

31. The value of science lies in its theoretical products.

32. Process skills are the most important things to be developed by children in science.

33. A major purpose of science is to produce new drugs and save lives.

34. I like science, and I probably will be (am) a better science teacher than most other teachers.

35. Science is devoted to describing how things happen.

36. I am afraid to teach science because I can't do the experiments myself.

37. Public understanding of science is necessary because scientific research requires financial support through the government.

38. I just never will understand science.

39. People need to understand the nature of science because it has such a great affect upon their lives.

40. A teacher has a responsibility to teach the basic facts of science.

41. Scientists discover laws which tell us exactly what is going on in nature.

42. The idea of teaching science scares me.

43. Demonstrations should be used frequently so the children will understand what their teacher tells them.

44. Scientists believe that they can find explanations for what they observe by looking at natural phenomena.

45. Scientific laws cannot be changed.

46. If an experiment does not come out right, the teacher should tell the children the answer so they will not be lost.

47. There are some things which are known by science to be absolutely true.

48. It is a teacher's responsibility to tell children which things are important for them to know.

49. I do (will) not teach very much science.

50. An important purpose of science is to help man to live longer.

51. A useful scientific theory may not be entirely correct, but it is the best idea scientists have been able to think up.

52. Today's electric appliances are examples of the really valuable products of science.

53. It is important for children to learn how to control variables in an experiment--at least by the sixth grade.

54. I am well prepared to teach science.

55. The teacher should arrange things so that children spend more time experimenting than listening to her in science.

56. Scientists are always interested in improving their explanations of natural events.

57. The value of science lies in its usefulness in solving practical problems.

58. I think I understand the nature of science and science teaching pretty well.

59. Most people are able to understand the work of science.

60. Scientific explanations can be made only by scientists.

61. Most children should know that the blood carries oxygen to the cells--at least by the sixth grade.

62. We can always get answers to our questions by asking a scientist.

63. Scientific laws have been proven beyond all possible doubt.

64. Looking at natural phenomena is a most important source of scientific information.

65. A major function of the teacher in teaching science is to help children identify problems.

66. If a scientist cannot answer a question, all he has to do is to ask another scientist.

67. Anything we need to know can be found out through science.

68. It is important for children to know why iron rusts--at least by the sixth grade.

69. Scientific ideas may be said to undergo a process of evolution in their development.

70. Scientists cannot always find the answers to their questions.

APPENDIX F

Perceptions of Communicator Attitudes (PCA)

Perceptions of Communicator Attitudes

<u>Instructions</u>: The purpose of this instrument is for you to record what <u>you think</u> the attitudes are of the following persons. Use Richard Moore's "What Is Your Attitude Toward Science And Science Teaching?" (STAS) instrument. As you read the statements, <u>ask yourself how your science instructor and graduate assistant would respond to the statements, then mark each statement</u> on the answer sheet by circling the letter of the appropriate response.

	SCIENCE INSTRUCTOR					GRADUATE ASSISTANT			
1.	A	B	C	D	1.	A	B	C	D
2.	A	B	C	D	2.	A	B	C	D
3.	A	B	C	D	3.	A	B	C	D
4.	A	B	C	D	4.	A	B	C	D
5.	A	B	C	D	5.	A	B	C	D
6.	A	B	C	D	6.	A	B	C	D
7.	A	B	C	D	7.	A	B	C	D
8.	A	B	C	D	8.	A	B	C	D
9.	A	B	C	D	9.	A	B	C	D
10.	A	B	C	D	10.	A	B	C	D
11.	A	B	C	D	11.	A	B	C	D
12.	A	B	C	D	12.	A	B	C	D
13.	A	B	C	D	13.	A	B	C	D
14.	A	B	C	D	14.	A	B	C	D
15.	A	B	C	D	15.	A	B	C	D
16.	A	B	C	D	16.	A	B	C	D
17.	A	B	C	D	17.	A	B	C	D
18.	A	B	C	D	18.	A	B	C	D
19.	A	B	C	D	19.	A	B	C	D
20.	A	B	C	D	20.	A	B	C	D
21.	A	B	C	D	21.	A	B	C	D
22.	A	B	C	D	22.	A	B	C	D
23.	A	B	C	D	23.	A	B	C	D
24.	A	B	C	D	24.	A	B	C	D
25.	A	B	C	D	25.	A	B	C	D
26.	A	B	C	D	26.	A	B	C	D
27.	A	B	C	D	27.	A	B	C	D
28.	A	B	C	D	28.	A	B	C	D
29.	A	B	C	D	29.	A	B	C	D
30.	A	B	C	D	30.	A	B	C	D
31.	A	B	C	D	31.	A	B	C	D
32.	A	B	C	D	32.	A	B	C	D
33.	A	B	C	D	33.	A	B	C	D
34.	A	B	C	D	34.	A	B	C	D
35.	A	B	C	D	35.	A	B	C	D

Perceptions of Communicator Attitudes
Page 2

	SCIENCE INSTRUCTOR					GRADUATE ASSISTANT			
36.	A	B	C	D	36.	A	B	C	D
37.	A	B	C	D	37.	A	B	C	D
38.	A	B	C	D	38.	A	B	C	D
39.	A	B	C	D	39.	A	B	C	D
40.	A	B	C	D	40.	A	B	C	D
41.	A	B	C	D	41.	A	B	C	D
42.	A	B	C	D	42.	A	B	C	D
43.	A	B	C	D	43.	A	B	C	D
44.	A	B	C	D	44.	A	B	C	D
45.	A	B	C	D	45.	A	B	C	D
46.	A	B	C	D	46.	A	B	C	D
47.	A	B	C	D	47.	A	B	C	D
48.	A	B	C	D	48.	A	B	C	D
49.	A	B	C	D	49.	A	B	C	D
50.	A	B	C	D	50.	A	B	C	D
51.	A	B	C	D	51.	A	B	C	D
52.	A	B	C	D	52.	A	B	C	D
53.	A	B	C	D	53.	A	B	C	D
54.	A	B	C	D	54.	A	B	C	D
55.	A	B	C	D	55.	A	B	C	D
56.	A	B	C	D	56.	A	B	C	D
57.	A	B	C	D	57.	A	B	C	D
58.	A	B	C	D	58.	A	B	C	D
59.	A	B	C	D	59.	A	B	C	D
60.	A	B	C	D	60.	A	B	C	D
61.	A	B	C	D	61.	A	B	C	D
62.	A	B	C	D	62.	A	B	C	D
63.	A	B	C	D	63.	A	B	C	D
64.	A	B	C	D	64.	A	B	C	D
65.	A	B	C	D	65.	A	B	C	D
66.	A	B	C	D	66.	A	B	C	D
67.	A	B	C	D	67.	A	B	C	D
68.	A	B	C	D	68.	A	B	C	D
69.	A	B	C	D	69.	A	B	C	D
70.	A	B	C	D	70.	A	B	C	D

Perceptions of Communicator Attitudes

Instructions: The purpose of this instrument is for you to record what <u>you think</u> the attitudes are of the following persons. Use Richard Moore's "What Is Your Attitude Toward Science And Science Teaching?" (STAS) instrument. As you read the statements, <u>ask yourself how your university supervisor, cooperating teacher, and peer member would respond to the statements, then mark each statement</u> on the answer sheet by circling the letter of the appropriate response.

	UNIVERSITY SUPERVISOR				COOPERATING TEACHER				PEER TEAM MEMBER			
1.	A	B	C	D	1. A	B	C	D	1. A	B	C	D
2.	A	B	C	D	2. A	B	C	D	2. A	B	C	D
3.	A	B	C	D	3. A	B	C	D	3. A	B	C	D
4.	A	B	C	D	4. A	B	C	D	4. A	B	C	D
5.	A	B	C	D	5. A	B	C	D	5. A	B	C	D
6.	A	B	C	D	6. A	B	C	D	6. A	B	C	D
7.	A	B	C	D	7. A	B	C	D	7. A	B	C	D
8.	A	B	C	D	8. A	B	C	D	8. A	B	C	D
9.	A	B	C	D	9. A	B	C	D	9. A	B	C	D
10.	A	B	C	D	10. A	B	C	D	10. A	B	C	D
11.	A	B	C	D	11. A	B	C	D	11. A	B	C	D
12.	A	B	C	D	12. A	B	C	D	12. A	B	C	D
13.	A	B	C	D	13. A	B	C	D	13. A	B	C	D
14.	A	B	C	D	14. A	B	C	D	14. A	B	C	D
15.	A	B	C	D	15. A	B	C	D	15. A	B	C	D
16.	A	B	C	D	16. A	B	C	D	16. A	B	C	D
17.	A	B	C	D	17. A	B	C	D	17. A	B	C	D
18.	A	B	C	D	18. A	B	C	D	18. A	B	C	D
19.	A	B	C	D	19. A	B	C	D	19. A	B	C	D
20.	A	B	C	D	20. A	B	C	D	20. A	B	C	D
21.	A	B	C	D	21. A	B	C	D	21. A	B	C	D
22.	A	B	C	D	22. A	B	C	D	22. A	B	C	D
23.	A	B	C	D	23. A	B	C	D	23. A	B	C	D
24.	A	B	C	D	24. A	B	C	D	24. A	B	C	D
25.	A	B	C	D	25. A	B	C	D	25. A	B	C	D
26.	A	B	C	D	26. A	B	C	D	26. A	B	C	D
27.	A	B	C	D	27. A	B	C	D	27. A	B	C	D
28.	A	B	C	D	29. A	B	C	D	29. A	B	C	D
30.	A	B	C	D	30. A	B	C	D	30. A	B	C	D
31.	A	B	C	D	31. A	B	C	D	31. A	B	C	D
32.	A	B	C	D	32. A	B	C	D	33. A	B	C	D
33.	A	B	C	D	33. A	B	C	D	33. A	B	C	D
34.	A	B	C	D	34. A	B	C	D	34. A	B	C	D
35.	A	B	C	D	35. A	B	C	D	35. A	B	C	D

Perceptions of Communicator Attitudes
Page 2

UNIVERSITY SUPERVISOR	COOPERATING TEACHER	PEER TEAM MEMBER
34. A B C D	34. A B C D	34. A B C D
35. A B C D	35. A B C D	35. A B C D
36. A B C D	36. A B C D	36. A B C D
37. A B C D	37. A B C D	37. A B C D
38. A B C D	38. A B C D	38. A B C D
40. A B C D	40. A B C D	40. A B C D
41. A B C D	41. A B C D	41. A B C D
42. A B C D	42. A B C D	42. A B C D
43. A B C D	43. A B C D	43. A B C D
44. A B C D	44. A B C D	44. A B C D
45. A B C D	45. A B C D	45. A B C D
46. A B C D	46. A B C D	46. A B C D
47. A B C D	47. A B C D	47. A B C D
48. A B C D	48. A B C D	48. A B C D
49. A B C D	49. A B C D	49. A B C D
50. A B C D	50. A B C D	50. A B C D
51. A B C D	51. A B C D	51. A B C D
52. A B C D	52. A B C D	52. A B C D
53. A B C D	53. A B C D	53. A B C D
54. A B C D	54. A B C D	54. A B C D
55. A B C D	55. A B C D	55. A B C D
56. A B C D	56. A B C D	56. A B C D
57. A B C D	57. A B C D	57. A B C D
58. A B C D	58. A B C D	58. A B C D
59. A B C D	59. A B C D	59. A B C D
60. A B C D	60. A B C D	60. A B C D
61. A B C D	61. A B C D	61. A B C D
62. A B C D	62. A B C D	62. A B C D
63. A B C D	63. A B C D	63. A B C D
64. A B C D	64. A B C D	64. A B C D
65. A B C D	65 A B C D	65. A B C D
66. A B C D	66. A B C D	66. A B C D
67. A B C D	67. A B C D	67. A B C D
68. A B C D	68. A B C D	68. A B C D
69. A B C D	69. A B C D	69. A B C D
70. A B C D	70. A B C D	70. A B C D

Appendix G

Perception of Communicator Credibility

Perception of Communicator Credibility

Instructions

The purpose of this instrument is to measure meanings of certain things to various people by having them judge them against a series of descriptive scales. In completing this instrument, please make your judgments on the basis of what these things mean to you.

Here is how you are to use these scales: If you feel that the concept at the top of the scales is very closely related to one end of the scale (for instance, very fair), you should place your mark as follows:

Fair _X_ : ___ : ___ : ___ : ___ : ___ : ___ Unfair

If you feel that the concept is only slightly related to one or the other end of the scale (for instance, slightly strong), you should place your mark as follows:

Weak ___ : ___ : ___ : ___ : _X_ : ___ : ___ Strong

The direction toward which you place your mark depends on which of the two ends of the scale seem most characteristic of the thing you are judging.

If you consider the concept to be neutral on the scale, i.e., both sides of the scale equally associated with the concept, or if the scale is completely unrelated to the concept, then place your mark in the middle space.

Rate the concept on each of the following scales. Mark them in order and do not omit any of the scales. Please do not look back and forth through the items and do not try to remember how you checked similar items earlier in the instrument. Make each item a separate and independent judgment. Work at fairly high speed and do not worry or puzzle over individual items. It is your first impression, your "immediate feelings" about the items which are desired. But, at the same time, please do not be careless, because your true impressions are wanted.

Unit Evaluator's <u>EXPERTNESS</u> in science and science teaching.

Bad	___	: ___	: ___	: ___	: ___	: ___	: ___	Good
Positive	___	: ___	: ___	: ___	: ___	: ___	: ___	Negative
Wise	___	: ___	: ___	: ___	: ___	: ___	: ___	Foolish
Weak	___	: ___	: ___	: ___	: ___	: ___	: ___	Strong
Active	___	: ___	: ___	: ___	: ___	: ___	: ___	Passive
Unpredictable	___	: ___	: ___	: ___	: ___	: ___	: ___	Predictable
Responsible	___	: ___	: ___	: ___	: ___	: ___	: ___	Irresponsible
Meaningless	___	: ___	: ___	: ___	: ___	: ___	: ___	Meaningful
Reputable	___	: ___	: ___	: ___	: ___	: ___	: ___	Disreputable

Unit Evaluator's <u>TRUSTWORTHINESS</u> in science and science teaching.

Predictable	___	: ___	: ___	: ___	: ___	: ___	: ___	Unpredictable
Foolish	___	: ___	: ___	: ___	: ___	: ___	: ___	Wise
Meaningful	___	: ___	: ___	: ___	: ___	: ___	: ___	Meaningless
Negative	___	: ___	: ___	: ___	: ___	: ___	: ___	Positive
Strong	___	: ___	: ___	: ___	: ___	: ___	: ___	Weak
Disreputable	___	: ___	: ___	: ___	: ___	: ___	: ___	Reputable
Good	___	: ___	: ___	: ___	: ___	: ___	: ___	Bad
Irresponsible	___	: ___	: ___	: ___	: ___	: ___	: ___	Responsible
Active	___	: ___	: ___	: ___	: ___	: ___	: ___	Passive

Graduate Assistant's EXPERTNESS in science and science teaching.

```
   Reputable ___ : ___ : ___ : ___ : ___ : ___ : ___ Disreputable
     Passive ___ : ___ : ___ : ___ : ___ : ___ : ___ Active
      Strong ___ : ___ : ___ : ___ : ___ : ___ : ___ Weak
         Bad ___ : ___ : ___ : ___ : ___ : ___ : ___ Good
 Responsible ___ : ___ : ___ : ___ : ___ : ___ : ___ Irresponsible
     Foolish ___ : ___ : ___ : ___ : ___ : ___ : ___ Wise
  Meaningful ___ : ___ : ___ : ___ : ___ : ___ : ___ Meaningless
    Negative ___ : ___ : ___ : ___ : ___ : ___ : ___ Positive
 Predictable ___ : ___ : ___ : ___ : ___ : ___ : ___ Unpredictable
```

Graduate Assistant's TRUSTWORTHINESS in science and science teaching.

```
          Bad ___ : ___ : ___ : ___ : ___ : ___ : ___ Good
     Positive ___ : ___ : ___ : ___ : ___ : ___ : ___ Negative
         Wise ___ : ___ : ___ : ___ : ___ : ___ : ___ Foolish
         Weak ___ : ___ : ___ : ___ : ___ : ___ : ___ Strong
       Active ___ : ___ : ___ : ___ : ___ : ___ : ___ Passive
Unpredictable ___ : ___ : ___ : ___ : ___ : ___ : ___ Predictable
  Responsible ___ : ___ : ___ : ___ : ___ : ___ : ___ Irresponsible
  Meaningless ___ : ___ : ___ : ___ : ___ : ___ : ___ Meaningful
    Reputable ___ : ___ : ___ : ___ : ___ : ___ : ___ Disreputable
```

University Supervisor's EXPERTNESS in science and science teaching.

Predictable ___ : ___ : ___ : ___ : ___ : ___ : ___ Unpredictable
Foolish ___ : ___ : ___ : ___ : ___ : ___ : ___ Wise
Meaningful ___ : ___ : ___ : ___ : ___ : ___ : ___ Meaningless
Negative ___ : ___ : ___ : ___ : ___ : ___ : ___ Positive
Strong ___ : ___ : ___ : ___ : ___ : ___ : ___ Weak
Disreputable ___ : ___ : ___ : ___ : ___ : ___ : ___ Reputable
Good ___ : ___ : ___ : ___ : ___ : ___ : ___ Bad
Irresponsible ___ : ___ : ___ : ___ : ___ : ___ : ___ Responsible
Active ___ : ___ : ___ : ___ : ___ : ___ : ___ Passive

University Supervisor's TRUSTWORTHINESS in science and science teaching.

Reputable ___ : ___ : ___ : ___ : ___ : ___ : ___ Disreputable
Passive ___ : ___ : ___ : ___ : ___ : ___ : ___ Active
Strong ___ : ___ : ___ : ___ : ___ : ___ : ___ Weak
Bad ___ : ___ : ___ : ___ : ___ : ___ : ___ Good
Responsible ___ : ___ : ___ : ___ : ___ : ___ : ___ Irresponsible
Foolish ___ : ___ : ___ : ___ : ___ : ___ : ___ Wise
Meaningful ___ : ___ : ___ : ___ : ___ : ___ : ___ Meaningless
Negative ___ : ___ : ___ : ___ : ___ : ___ : ___ Positive
Predictable ___ : ___ : ___ : ___ : ___ : ___ : ___ Unpredictable

Cooperating Teacher's EXPERTNESS in science and science teaching.

Bad	___ :	___ :	___ :	___ :	___ :	___	Good
Positive	___ :	___ :	___ :	___ :	___ :	___	Negative
Wise	___ :	___ :	___ :	___ :	___ :	___	Foolish
Weak	___ :	___ :	___ :	___ :	___ :	___	Strong
Active	___ :	___ :	___ :	___ :	___ :	___	Passive
Unpredictable	___ :	___ :	___ :	___ :	___ :	___	Predictable
Responsible	___ :	___ :	___ :	___ :	___ :	___	Irresponsible
Meaningless	___ :	___ :	___ :	___ :	___ :	___	Meaningful
Reputable	___ :	___ :	___ :	___ :	___ :	___	Disreputable

Cooperating Teacher's TRUSTWORTHINESS in science and science teaching.

Predictable	___ :	___ :	___ :	___ :	___ :	___	Unpredictable
Foolish	___ :	___ :	___ :	___ :	___ :	___	Wise
Meaningful	___ :	___ :	___ :	___ :	___ :	___	Meaningless
Negative	___ :	___ :	___ :	___ :	___ :	___	Positive
Strong	___ :	___ :	___ :	___ :	___ :	___	Weak
Disreputable	___ :	___ :	___ :	___ :	___ :	___	Reputable
Good	___ :	___ :	___ :	___ :	___ :	___	Bad
Irresponsible	___ :	___ :	___ :	___ :	___ :	___	Responsible
Active	___ :	___ :	___ :	___ :	___ :	___	Passive

Science Instructor's <u>EXPERTNESS</u> in science and science teaching.

Reputable	___ :	___ :	___ :	___ :	___ :	___ :	___ Disreputable
Passive	___ :	___ :	___ :	___ :	___ :	___ :	___ Active
Strong	___ :	___ :	___ :	___ :	___ :	___ :	___ Weak
Bad	___ :	___ :	___ :	___ :	___ :	___ :	___ Good
Responsible	___ :	___ :	___ :	___ :	___ :	___ :	___ Irresponsible
Foolish	___ :	___ :	___ :	___ :	___ :	___ :	___ Wise
Meaningful	___ :	___ :	___ :	___ :	___ :	___ :	___ Meaningless
Negative	___ :	___ :	___ :	___ :	___ :	___ :	___ Positive
Predictable	___ :	___ :	___ :	___ :	___ :	___ :	___ Unpredictable

Science Instructor's <u>TRUSTWORTHINESS</u> in science and science teaching.

Bad	___ :	___ :	___ :	___ :	___ :	___ :	___ Good
Positive	___ :	___ :	___ :	___ :	___ :	___ :	___ Negative
Wise	___ :	___ :	___ :	___ :	___ :	___ :	___ Foolish
Weak	___ :	___ :	___ :	___ :	___ :	___ :	___ Strong
Active	___ :	___ :	___ :	___ :	___ :	___ :	___ Passive
Unpredictable	___ :	___ :	___ :	___ :	___ :	___ :	___ Predictable
Responsible	___ :	___ :	___ :	___ :	___ :	___ :	___ Irresponsible
Meaningless	___ :	___ :	___ :	___ :	___ :	___ :	___ Meaningful
Reputable	___ :	___ :	___ :	___ :	___ :	___ :	___ Disreputable

Peer Team Member's EXPERTNESS in science and science teaching.

Predictable ___ : ___ : ___ : ___ : ___ : ___ : ___ Unpredictable
Foolish ___ : ___ : ___ : ___ : ___ : ___ : ___ Wise
Meaningful ___ : ___ : ___ : ___ : ___ : ___ : ___ Meaningless
Negative ___ : ___ : ___ : ___ : ___ : ___ : ___ Positive
Strong ___ : ___ : ___ : ___ : ___ : ___ : ___ Weak
Disreputable ___ : ___ : ___ : ___ : ___ : ___ : ___ Reputable
Good ___ : ___ : ___ : ___ : ___ : ___ : ___ Bad
Irresponsible ___ : ___ : ___ : ___ : ___ : ___ : ___ Responsible
Active ___ : ___ : ___ : ___ : ___ : ___ : ___ Passive

Peer Team Member's TRUSTWORTHINESS in science and science teaching.

Reputable ___ : ___ : ___ : ___ : ___ : ___ : ___ Disreputable
Passive ___ : ___ : ___ : ___ : ___ : ___ : ___ Active
Strong ___ : ___ : ___ : ___ : ___ : ___ : ___ Weak
Bad ___ : ___ : ___ : ___ : ___ : ___ : ___ Good
Responsible ___ : ___ : ___ : ___ : ___ : ___ : ___ Irresponsible
Foolish ___ : ___ : ___ : ___ : ___ : ___ : ___ Wise
Meaningful ___ : ___ : ___ : ___ : ___ : ___ : ___ Meaningless
Negative ___ : ___ : ___ : ___ : ___ : ___ : ___ Positive
Predictable ___ : ___ : ___ : ___ : ___ : ___ : ___ Unpredictable

www.ingramcontent.com/pod-product-compliance
Lightning Source LLC
Chambersburg PA
CBHW051100230426
43667CB00013B/2373